Advances in Sustainable Concrete for Construction

Alireza Bahrami
Editor

Advances in Sustainable Concrete for Construction

Springer

Editor
Alireza Bahrami
Department of Building Engineering, Energy Systems
and Sustainability Science, Faculty of Engineering and Sustainable Development
University of Gävle
Gävle, Sweden

ISBN 978-3-031-85051-6 ISBN 978-3-031-85052-3 (eBook)
https://doi.org/10.1007/978-3-031-85052-3

This Springer imprint is published by the registered company Springer Nature Switzerland AG
The registered company address is: Gewerbestrasse 11, 6330 Cham, Switzerland

This book is dedicated to my beloved wife and dear son

Preface

Concrete is essential to modern construction, shaping the infrastructure that forms the foundation of our daily lives. However, its environmental impact, particularly due to cement production and the carbon emissions associated with it, poses a major challenge as we aim for more sustainable development. This book delves into innovative solutions to enhance concrete's sustainability, durability, and overall environmental impact.

One of the key ways to reduce concrete's carbon footprint is to use less cement by replacing it with industrial by-products like fly ash and slag. While these alternatives can significantly reduce emissions, it is crucial to strike a careful balance in their use to ensure the concrete continues to meet necessary performance standards. Structural optimization, particularly through more efficient reinforced concrete designs, also offers a way to reduce material use without compromising safety or stability.

As we move toward more sustainable infrastructure, advancements in concrete technology are indispensable. Alternative materials such as supplementary cementitious materials and carbon capture and storage solutions provide effective pathways to reduce emissions while maintaining the strength and durability of concrete. Additionally, life cycle assessments and economic considerations for these modern materials are vital in balancing environmental impact and cost-effectiveness. Predictive models developed through non-destructive testing methods provide practical tools for assessing the long-term performance of concrete structures, helping us make informed decisions about their lifespan and sustainability.

Emerging materials, such as engineered cementitious composites, are pushing the boundaries of what is possible in construction. These composites offer not only enhanced tensile strength and resilience but also contribute to sustainability by reducing material consumption and extending the life of structures. Techniques that enhance durability, particularly in harsh environments, are also making a notable difference. The incorporation of rice husk ash and hybrid fibers, such as alkali-resistant glass and steel fibers, provides effective solutions for projects exposed to chemically aggressive conditions, improving concrete's resistance to acid attacks.

Internal curing methods are another exciting innovation that can transform modern concrete practices. By sustaining hydration and reducing shrinkage, these methods improve the mechanical properties and overall longevity of concrete, addressing some of the traditional challenges with curing. Moreover, incorporating waste materials like spent wash into construction processes demonstrates how resource efficiency can combine to create more sustainable approaches.

Finally, the practice of earth construction, long used in traditional building techniques, is receiving renewed attention for its potential as a sustainable building alternative. Self-consolidating earth concrete offers a promising way forward, overcoming many of the limitations of traditional methods while maintaining the environmental and thermal benefits of earth. By optimizing its workability and strength, earth concrete holds great potential for contemporary applications.

This book presents the continuous effort to find sustainable solutions that bridge the gap between construction practices and environmental responsibility. It aims to inspire researchers, engineers, and industry professionals to continue exploring novel solutions in sustainable concrete technologies, shaping a future where infrastructure development and ecological responsibility are seamlessly integrated.

Gävle, Sweden Alireza Bahrami
2025

Contents

Structural Performance of Various Climate-Improved Concrete Types

Alireza Bahrami and Ahmed Shurie

1 Introduction

Concrete is a composite building material that has long been a cornerstone of construction thanks to its exceptional strength, durability, low maintenance, long lifespan, and versatility [1]. It is extensively employed in various construction projects, composed of a mixture of cement, aggregates, water, and additives [2]. As the most consumed construction material worldwide, with an annual production of approximately 30 billion tons, concrete plays a pivotal role in the construction industry, finding applications in essential structures such as bridges, buildings, and tunnels [3]. Despite its widespread use, the production of concrete has significant environmental implications, primarily due to its reliance on cement. Cement production involves the calcination of limestone, which releases large amounts of carbon dioxide—a key contributor to global greenhouse gas emissions. Cement is responsible for a substantial portion of the carbon dioxide emissions associated with concrete, which makes addressing these emissions a critical aspect of global climate mitigation efforts. Concrete production noticeably contributes to carbon dioxide emissions and the gradual depletion of natural resources, representing about 10% of global greenhouse gas emissions and 30% of the consumption of nonrenewable natural resources [4].

Growing concerns about sustainability and the increased emphasis on using renewable and recyclable materials have gained significant attention in recent times. The construction and real estate sector, being a substantial consumer of natural resources, contributes to approximately 21% of Sweden's total carbon dioxide emissions [5]. Encouragingly, the construction industry is making strides toward

A. Bahrami (✉) · A. Shurie
Department of Building Engineering, Energy Systems and Sustainability Science, Faculty of Engineering and Sustainable Development, University of Gävle, Gävle, Sweden
e-mail: alireza.bahrami@hig.se

A. Bahrami (ed.), *Advances in Sustainable Concrete for Construction*,
https://doi.org/10.1007/978-3-031-85052-3_1

increased sustainability, despite encountering numerous conflicting objectives and intricate challenges. Recognizing the urgency to tackle these issues, the industry has witnessed the emergence of alternative sustainable solutions aimed at assisting in the attainment of climate targets.

Globally, cement production contributes to approximately 3–4% of total carbon dioxide emissions, with the primary sources being the heating of limestone and the use of energy, primarily derived from the combustion of fossil fuels [6].

In Sweden, where ambitious net-zero carbon dioxide targets have been set, reducing the carbon footprint of concrete production is an essential part of the strategy to meet these goals. One promising approach to achieving this reduction is the use of climate-improved concrete. This form of concrete substitutes a portion of the traditional cement content with industrial by-products or waste materials, such as fly ash, slag, and recycled aggregates. By lowering the cement content, the carbon emissions from concrete production can be remarkably reduced without compromising the structural integrity of the material. This method also presents an opportunity to recycle industrial waste, thus contributing to a more circular economy within the construction industry.

In pursuit of its climate objectives, Sweden implemented a climate policy framework in 2017, with the ambitious goal of achieving net-zero greenhouse gas emissions by 2045 [7]. To address the challenge of reducing carbon dioxide emissions associated with concrete production, the concept of climate-improved concrete has emerged as a promising solution. Climate-improved concrete shares fundamental similarities with conventional concrete, consisting of water, cement, aggregates, and additives. However, what sets climate-improved concrete apart is the deliberate substitution of a portion of cement with industrial waste materials such as fly ash and blast furnace slag. These alternative materials exhibit lower carbon dioxide emissions compared to cement. Moreover, the replacement of natural aggregates in the concrete mixture by sustainable alternatives such as recycled concrete aggregates, glass, and demolition waste further enhances its climate-improved characteristics [8]. Through these innovative approaches, climate-improved concrete aims to mitigate the environmental impact of concrete production and contribute to the achievement of sustainable construction practices.

These issues have led to a research focus for engineers who are exploring potential alternatives for producing climate-improved concrete [9]. Advancements in concrete technology have been steadily growing due to numerous studies [10, 11]. However, this chapter involves a comprehensive literature review on the effects of using industrial by-products in concrete on various concrete properties, including the compressive strength, tensile strength, modulus of elasticity, carbonation depth, creep, workability, and drying shrinkage.

2 Methodology

The methodology employed in this study includes a comprehensive literature review on climate-improved concrete to gather data. The literature review encompassed an examination of scientific articles pertaining to climate-improved concrete and the utilization of cementitious materials. The selection of relevant articles was executed through the use of search portals and reputable scientific journals including ScienceDirect.

Multiple scientific articles were diligently reviewed, with a subset of 19 articles being selected for comparative analysis in this study. Each article was assigned a distinct number in Table 1, delineating the authors and the specific concrete properties investigated.

Concrete mixtures that demonstrated optimal properties, as indicated by the selected articles, were meticulously selected, categorized, and grouped based on the replacement materials employed. The comparison of different types of climate-improved concrete focused on the individual examination of concrete properties. The structural performance of these concrete mixtures was comprehensively analyzed, and the findings are presented and discussed in the following section.

Table 1 Investigated properties of concrete mixtures in reviewed articles

Article no.	Author	Investigated concrete properties[a]
1	Yüksel et al. [12]	CS
2	Domingo-Cabo et al. [13]	CS, DS, C, ME, W
3	Corinaldesi and Moriconi [14]	CS, TS, DS, ME, CD, W
4	Corinaldesi and Moriconi [15]	CS, DS, W
5	Thomas et al. [16]	CS, TS, ME, CD
6	Alwaeli [17]	CS
7	Radonjanin et al. [18]	CS, TS, ME, W
8	Van Dr. Wegen et al. [19]	CS, TS, ME, DS, C, CD, W
9	Lima et al. [20]	CS, TS, W
10	Proske et al. [21]	CS, TS, ME, DS
11	Valcuende et al. [22]	CS, ME, DS
12	Turk et al. [23]	CS, ME, W
13	Rondi et al. [24]	CS, DS
14	Faella et al. [25]	CS, TS, CD
15	Gonzalez-Corominas and Etxeberria [26]	CS, ME, DS
16	Golewski [27]	CS
17	Marinković et al. [28]	CS, W
18	Velay-Lizancos et al. [29]	CS
19	Gebremariam et al. [30]	CS, ME, CD, W

[a]Abbreviations: *CS* compressive strength, *TS* tensile strength, *ME* modulus of elasticity, *DS* drying shrinkage; *C* creep, *CD* carbonation depth, *W* workability

3 Results and Discussion

This study aimed to compare the effects of alternative materials on concrete properties for reducing carbon dioxide emissions. Previous research has shown that using fly ash or slag as partial substitutes for cement enhances concrete strength during longer curing times but slows down strength development. Conversely, the inclusion of recycled aggregates in high volumes diminishes concrete strength. The results are explained and discussed based on the investigated concrete properties.

3.1 Compressive Strength

The compressive strength, being a fundamental characteristic of concrete, was consistently addressed across all the referenced articles. To explore the impact of alternative replacement materials on concrete properties, specific attention was directed toward concrete mixtures featuring fly ash, blast furnace slag, and recycled aggregates. These mixtures were categorized into distinct groups based on the replacement materials used in the concrete composition.

3.1.1 Fly Ash or Bottom Ash

Detailed information regarding the concrete mixtures containing fly ash or bottom ash, including their designations, replacement materials, and the primary source from which they originated, can be found in Table 2. The assessment of 28-day

Table 2 Concrete mixtures containing fly ash or bottom ash

Article no.	Concrete mixture	Explanation
1	CK-5	Concrete mixture containing 5% granulated blast furnace slag and 5% bottom ash
3	REC + FA	Concrete mixture containing 100% recycled aggregates and 38% fly ash
8	G20	Concrete mixture where 20% of gravel was replaced by bottom ash
9	MR30	Concrete mixture containing 30% recycled aggregates and 47% fly ash
10	ECO-3	Concrete mixture containing limestone powder and 50% fly ash
12	G-FA	Concrete mixture containing fly ash
14	MR30-A	Concrete mixture containing 30% recycled aggregates and 47% fly ash
16	CFA-20	Concrete mixture containing 20% coal fly ash
17	NAC_AAFA	Concrete mixture containing alkaline-activated fly ash and 100% natural aggregates
	RAC_FA	Concrete mixture containing fly ash and 100% recycled aggregates

compressive strength was conducted for the selected concrete mixtures incorporating fly ash or bottom ash as a substitute material, and the corresponding findings are illustrated in Fig. 1.

The comparative analysis reveals that the CK-5, ECO-3, and CFA-20 mixtures displayed superior compressive strength values in comparison to the remaining mixtures. Upon assessing the concrete mixtures depicted in Fig. 1, it was evident that those comprising solely of 100% natural aggregates provided higher compressive strength values than their counterparts containing recycled aggregates. Furthermore, the concrete mixtures composed solely of natural aggregates achieved the compressive strength values exceeding 40 MPa.

Further investigation into the mixtures consisting of 100% natural aggregates revealed that the top performing mixtures were formulated with CEM I type of cement, featuring relatively lower water-to-cement (w/c) values and a reduced content of fly ash. This observation suggests that the utilization of such cement type, along with the decreased w/c ratio and fly ash content, likely contributed to the notable increase in the compressive strength values.

The comparative analysis indicates that the CK-5, ECO-3, and CFA-20 mixtures demonstrated superior compressive strength values compared to the remaining mixtures. However, those mixtures solely with 100% natural aggregates manifested higher compressive strength values compared to mixtures having recycled aggregates.

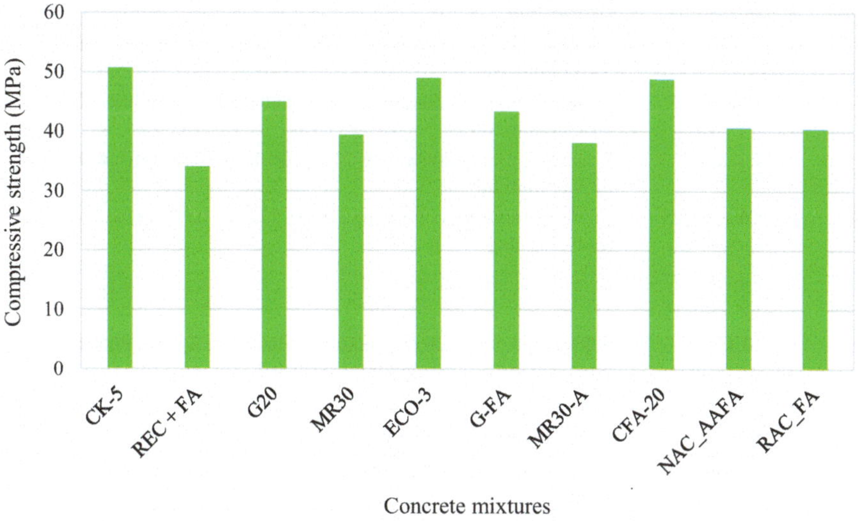

Fig. 1 Compressive strength of concrete mixtures containing fly ash or bottom ash

3.1.2 Slag

In the assessment of the compressive strength of the concrete mixtures incorporating slag, a meticulous assessment was conducted, and 13 concrete mixtures, as outlined in Table 3, were specifically selected for the analysis.

The findings concerning concrete mixtures having slag as a replacement material are shown in Fig. 2, where the CK-5 and GLZSC-100 mixtures exhibited the highest compressive strength values. These results mirror the trends observed in mixtures containing fly ash. The concrete mixtures containing slag revealed comparable high compressive strength values to those observed in mixtures containing fly ash.

However, the CK-5 and C25/30 mixtures deviated from the observed trend regarding the influence of slag content on the strength values. The compressive strength of the CK mixtures illustrated a gradual decline with increasing slag volume. In contrast, the GLZSC mixtures displayed an increase in compressive strength as the slag content rose. These contrasting outcomes can be attributed to the inclusion of fly ash, bottom ash, and granulated blast furnace slag in the CK mixtures. In addition, the elevated compressive strength witnessed in the GLZSC mixtures can be linked to the higher cement content employed.

With regard to the compressive strength, the self-compacting concrete mixtures (denoted by SCC in Table 3 and Fig. 2), excluding the C25/30 mixture, indicated lower strength values compared to the other concrete mixtures. This deviation can be due to the utilization of a combination of fly ash, slag, and limestone filler material in the self-compacting concrete mixtures.

Table 3 Concrete mixtures containing slag

Article no.	Concrete mixture	Explanation
1	CK-5	Concrete mixture containing 5% granulated blast furnace slag and 5% bottom ash
	CK-15	Concrete mixture containing 15% granulated blast furnace slag and 15% bottom ash
	CK-25	Concrete mixture containing 25% granulated blast furnace slag and 25% bottom ash
6	GLZSC-25	Concrete mixtures containing 25, 50, 75, and 100% granulated lead–zinc slag
	GLZSC-50	
	GLZSC-75	
	GLZSC-100	
11	SCC-20	Self-compacting concrete mixtures containing fly ash, limestone filler, and 20%, 40%, and 60% slag sand
	SCC-40	
	SCC-60	
12	G-SL	Concrete mixture containing slag
	G-REC-SL	Concrete mixture containing steel slag and 100% recycled aggregates
13	C25/30	Concrete mixture with strength class C25/30 and containing steel slag as aggregates

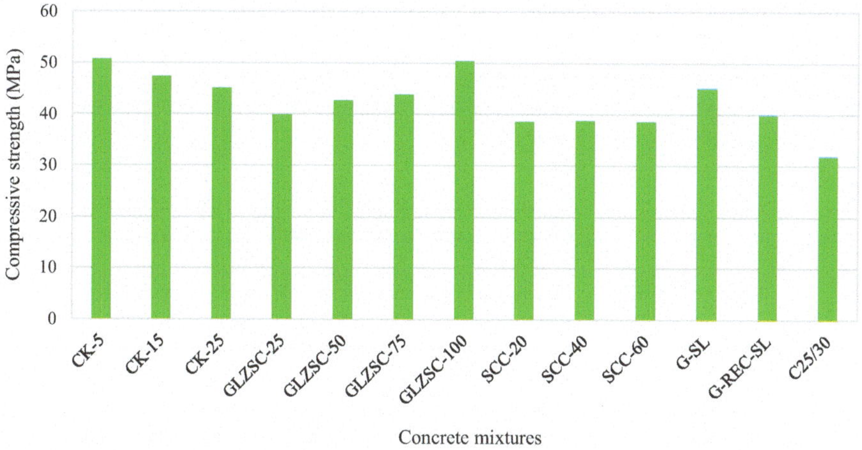

Fig. 2 Compressive strength of concrete mixtures containing slag

The reduced compressive strength value seen for the C25/30 mixture can likely be owing to its classification as conventional concrete, wherein the entire volume of natural aggregates was substituted with slag as an artificial aggregate. Moreover, it is plausible that the w/c ratio and the specific type or class of cement exerted a significant influence on the compressive strength outcomes of the concrete mixtures.

3.1.3 Recycled Aggregates and/or Other Additives

Table 4 lists information on 16 concrete mixtures that have recycled aggregates and/or other additives.

The findings depicted in Fig. 3 reveal that, in contrast to expectations, the concrete mixtures having recycled aggregates provided generally higher compressive strength compared to mixtures incorporating fly ash or slag. This observation contradicts the earlier observations made regarding the concrete mixtures combining fly ash with recycled aggregates. One potential explanation for this discrepancy is that the use of high-strength concrete as aggregates may have contributed to the overall increase in the compressive strength.

Further, Fig. 3 demonstrates a consistent distribution of the compressive strength values across the various concrete mixtures, with a significant majority of the selected mixtures surpassing the threshold of 40 MPa. The H40 mixtures showed an enhanced compressive strength as the proportion of recycled aggregates increased. Among these mixtures, the H40-C mixture provided the highest compressive strength in comparison to the others. In this particular mixture, 100% natural fine aggregates were employed, while all coarse aggregates were replaced by recycled aggregates. The remarkable increase in the compressive strength of the H40-C mixture, compared to the other mixtures, can be related to the greater absorption capacity of recycled aggregates, resulting in a reduced effective w/c ratio.

Table 4 Concrete mixtures containing recycled aggregates and/or other additives

Article no.	Concrete mixture	Explanation
2	H40-A	Concrete mixtures containing 20, 50, and 100% recycled aggregates
	H40-B	
	H40-C	
3	REC	Concrete mixture containing 100% recycled aggregates
	REC + FA	Concrete mixture containing 100% recycled aggregates and 38% fly ash
4	REC + SP + SRA	Concrete mixture containing 100% recycled aggregates, superplasticizer, and a shrinkage-reducing admixture
	REC + SP	Concrete mixture containing 100% recycled aggregates and superplasticizer
	RECSAT + SP + SRA	Concrete mixture containing 100% saturated recycled aggregates, superplasticizer, and a shrinkage-reducing admixture
5	H-0.45ARS-20	Concrete mixture with w/c ratio of 0.45 and containing 20%, 50%, and 100% recycled aggregates
	H-0.45ARS-50	
	H-0.45ARS-100	
7	A-SF	Concrete mixtures containing silica fume or metakaolin
	A-MK	
9	MR30	Concrete mixture containing 30% recycled aggregates and 47% fly ash
12	REC-A	Concrete mixture containing 100% recycled aggregates
	G-REC-SL	Concrete mixture containing 100% recycled aggregates and slag

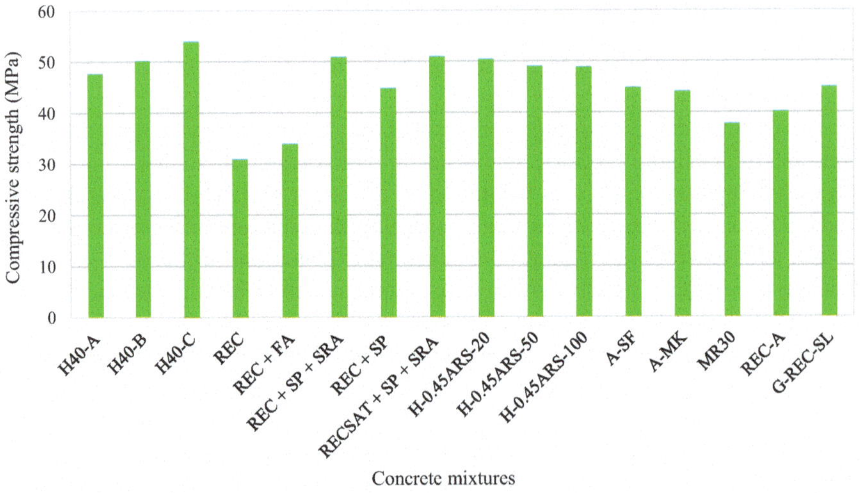

Fig. 3 Compressive strength of concrete mixtures containing recycled aggregates and/or other additives

Additionally, the effects of the additive used, as indicated in previous research [13], may have contributed to the observed outcome. Conversely, the H-0.45ARS, A-SF, A-MK, MR30, REC-A, and G-REC-SL mixtures exhibited lower compressive strength values compared to the H40-C mixture. This disparity can be because these mixtures had a lower cement content, coupled with varying percentages of recycled aggregates. The REC and REC + FA mixtures presented the lowest compressive strength values among all the other mixtures. A plausible explanation is that these mixtures were composed of 100% recycled aggregates and utilized the CEM II type of cement, which may include materials like fly ash, blast furnace slag, limestone, or silica fume.

The concrete mixtures having superplasticizers and recycled aggregates illustrated diverse compressive strength values. One interpretation of these variations is that the inclusion of recycled aggregates, in conjunction with the use of a shrinkage-reducing admixture, led to a certain degree of increase in the compressive strength.

3.2 *Tensile Strength*

In addition to compressive strength, tensile strength is also a crucial concrete property considered in the comparative assessment of climate-improved concrete mixtures conducted in this study. Concrete mixtures sourced from different articles were selected for the purpose of comparing their 28-day tensile strengths. To facilitate the analysis, the selected concrete mixtures were categorized into two groups based on the types of replacement materials utilized.

3.2.1 Fly Ash or Bottom Ash

The tensile strength analysis was done on 12 concrete mixtures selected for this study. Table 5 highlights an overview of the replacement materials used in these mixtures, along with their corresponding abbreviations. The table also references the original articles from which the concrete mixtures were obtained.

The comparative analysis of the tensile strength revealed a diverse distribution among the selected concrete mixtures. As depicted in Fig. 4, the ECO-2, ECO-3, and G20 mixtures exhibited the highest tensile strength values. The reasons underlying these mixtures' superior tensile strength values can be owing to several factors, including a high cement content, a minimal proportion of recycled aggregates, and relatively low levels of replacement materials such as fly ash or bottom ash.

The G20 mixture stood out with the highest cement content, coupled with a significant proportion of natural aggregates. On the other hand, the ECO-2 and ECO-3 mixtures demonstrated lower cement content but utilized 100% natural aggregates. In contrast, the remaining concrete mixtures featured reduced cement content, elevated amounts of recycled aggregates, and, in certain instances, substantial volumes of replacement materials, which had a detrimental impact on the tensile strength.

Table 5 Concrete mixtures containing fly ash or bottom ash

Article no.	Concrete mixture	Explanation
3	REC + FA	Concrete mixture containing 100% recycled aggregates and 38% fly ash
8	G20	Concrete mixture where 20% of gravel was replaced by bottom ash
	G20 + S20	Concrete mixture where 20% of gravel and 20% of sand were replaced by bottom ash
9	MN	Concrete mixture containing 100% natural aggregates and 47% fly ash
	MR30	Concrete mixture containing 30% recycled aggregates and 47% fly ash
	HN	Concrete mixture containing 100% natural aggregates and 56% fly ash
10	ECO-1	Concrete mixture containing limestone powder and 15% fly ash
	ECO-2	Concrete mixture containing limestone powder and 33% fly ash
	ECO-3	Concrete mixture containing limestone powder and 50% fly ash
14	LR30-A	Concrete mixture containing 30% recycled aggregates and 24% fly ash
	MR30-A	Concrete mixture containing 30% recycled aggregates and 47% fly ash
	HR30-A	Concrete mixture containing 30% recycled aggregates and 56% fly ash

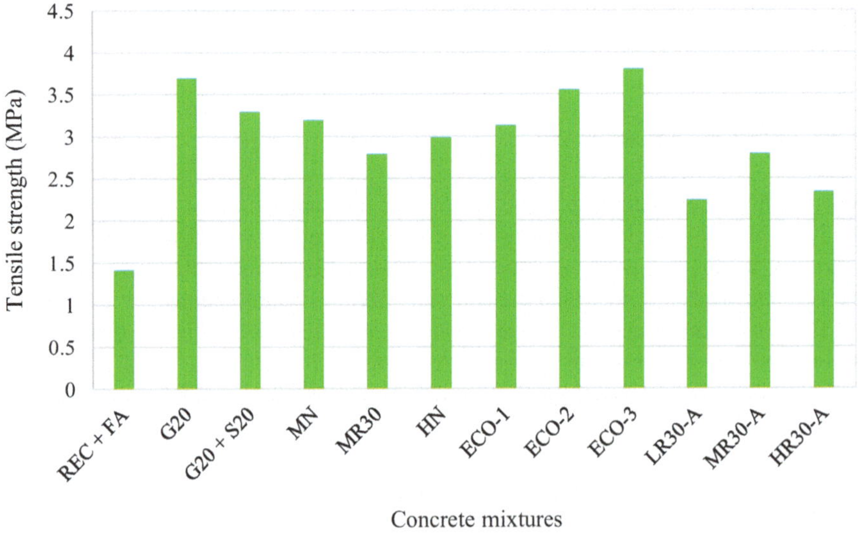

Fig. 4 Tensile strength of concrete mixtures containing fly ash or bottom ash

The REC + FA mixture displayed the lowest tensile strength when compared to the other mixtures. This can be due to the complete substitution of natural aggregates with recycled aggregates, as well as the high content of fly ash.

Concrete mixtures incorporating fly ash and bottom ash indicated a general improvement in the tensile strength. Fig. 4 shows two distinct groups of concrete mixtures. The first group comprises mixtures exclusively composed of natural aggregates, which provided commendable tensile strength values. An increasing volume of fly ash corresponded to higher tensile strength. The second group includes mixtures having recycled aggregates. It was observed that these mixtures presented lower tensile strength compared to those utilizing solely natural aggregates. A notable example is the REC + FA concrete mixture, which demonstrated the lowest strength value among the mixtures. However, the incorporation of fly ash in the concrete mixture mitigated the adverse effects stemming from the utilization of recycled concrete aggregates as substitute for natural aggregates.

3.2.2 Recycled Aggregates

In the evaluation of the tensile strength across the concrete mixtures incorporating recycled aggregates, a total of nine specific mixtures were selected for the analysis, as outlined in Table 6.

Concrete mixtures having recycled aggregates generally exhibited lower tensile strength compared to those incorporating fly ash as a replacement material. The comparison revealed that the H-0.45ARS mixtures had the highest tensile strength values, as depicted in Fig. 5. This can be related to their elevated cement content compared to the other concrete mixtures. Further, the H-0.45ARS mixtures featured a relatively high volume of natural aggregates. Fig. 5 also illustrates a clear decrease in the tensile strength as the proportion of recycled aggregates increased.

Table 6 Concrete mixtures containing recycled aggregates

Article no.	Concrete mixture	Explanation
3	REC	Concrete mixture containing 100% recycled aggregates
	REC + FA	Concrete mixture containing 100% recycled aggregates and 38% fly ash
5	H-0.45ARS-A	Concrete mixtures with a w/c ratio of 0.45 and containing 20, 50, and 100% recycled aggregates
	H-0.45ARS-B	
	H-0.45ARS-C	
9	MR30	Concrete mixture containing 30% recycled aggregates and 47% fly ash
14	LR30-A	Concrete mixture containing fly ash and recycled ballast
	MR30-A	Concrete mixture containing 30% recycled aggregates and 47% fly ash
	HR30-A	Concrete mixture containing 30% recycled aggregates and 56% fly ash

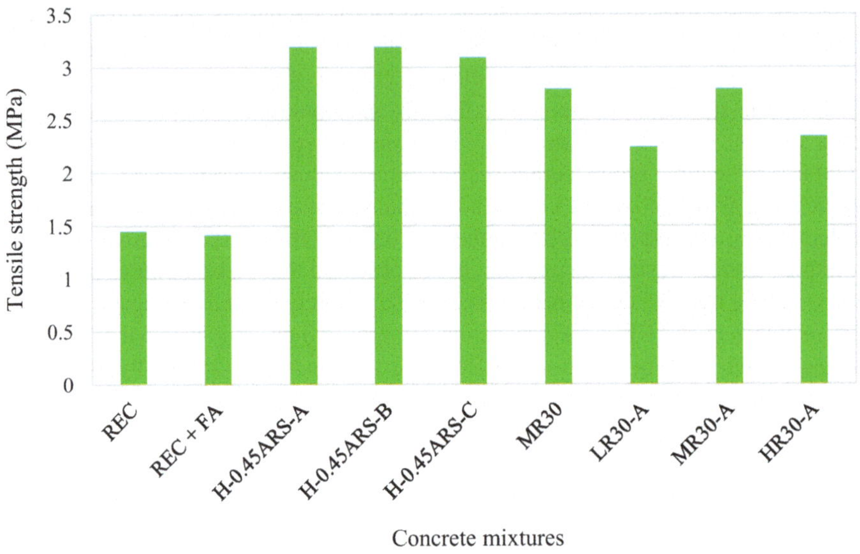

Fig. 5 Tensile strength of concrete mixtures containing recycled aggregates

A similar decreasing trend in the tensile strength was witnessed for the LR30-A, MR30-A, and HR30-A mixtures. The higher strength value of the MR30-A mixture compared to the LR30-A and HR30-A mixtures can be owing to its higher cement content than the HR30-A mixture, as well as the increased fly ash content in the MR30-A mixture compared to the LR30-A mixture. Moreover, the H-0.45ARS mixtures displayed a comparable decreasing trend, which was partially mitigated by the variation in the w/c ratio for each specific concrete mixture.

3.3 Drying Shrinkage

For the purpose of comparing the drying shrinkage, a total of 20 distinct concrete mixtures were carefully selected and categorized into different groups based on the specific replacement materials utilized. All the drying shrinkage values considered in the comparison were derived from the test results of the concrete mixtures following a 90-day curing period.

3.3.1 Fly Ash or Bottom Ash

The assessment of the drying shrinkage in the selected concrete mixtures focused specifically on the mixtures having fly ash or bottom ash, as outlined in Table 7.

Table 7 Concrete mixtures containing fly ash or bottom ash

Article no.	Concrete mixture	Explanation
3	REC + FA	Concrete mixture containing 100% recycled aggregates and 38% fly ash
8	G20	Concrete mixture where 20% of gravel was replaced by bottom ash
	G20 + S20	Concrete mixture where 20% of gravel and 20% of sand were replaced by bottom ash
10	ECO-1	Concrete mixture containing limestone powder and 15% fly ash
	ECO-2	Concrete mixture containing limestone powder and 33% fly ash
	ECO-3	Concrete mixture containing limestone powder and 50% fly ash

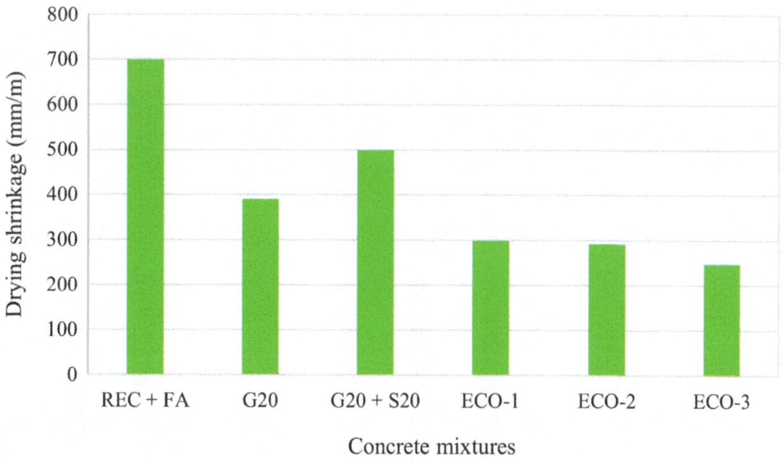

Fig. 6 Drying shrinkage of concrete mixtures containing fly ash or bottom ash

When evaluating the drying shrinkage of the concrete mixtures containing fly ash, the findings shown in Fig. 6 reveal significant insights. The ECO mixtures indicated the lowest drying shrinkage values compared to the other mixtures. This favorable outcome can be partially due to the relatively low fly ash content and the absence of recycled aggregates in the concrete composition. However, the G20 and G20 + S20 mixtures demonstrated slightly higher drying shrinkage values, which can be because of the increased ash content present in these mixtures.

Remarkably, the REC + FA mixture emerged with the highest drying shrinkage among the analyzed mixtures. This distinctive behavior can be owing to the substantial fly ash content and the complete substitution of natural aggregates with recycled aggregates in this particular mixture. The increased drying shrinkage is closely linked to the higher volume of micropores in the concrete's pore structure, as suggested by the findings of previous research [14].

Table 8 Concrete mixtures containing recycled aggregates

Article no.	Concrete mixture	Explanation
2	H40 B-A	Concrete mixtures containing 20, 50, and 100% recycled aggregates
	H40 B-B	
	H40 B-C	
3	REC	Concrete mixture containing 100% recycled aggregates
	REC + FA	Concrete mixture containing 100% recycled aggregates and 38% fly ash
4	REC + SP + SRA	Concrete mixture containing 100% recycled aggregates, superplasticizer, and shrinkage-reducing admixture
	REC + SP	Concrete mixture containing 100% recycled aggregates and superplasticizer
	RECSAT + SP + SRA	Concrete mixture containing 100% saturated recycled aggregates, superplasticizer, and shrinkage-reducing admixture
15	R40–100	Concrete mixtures containing 100% recycled aggregates with compressive strengths of 40, 60, and 100 MPa
	R60–100	
	R100–100	

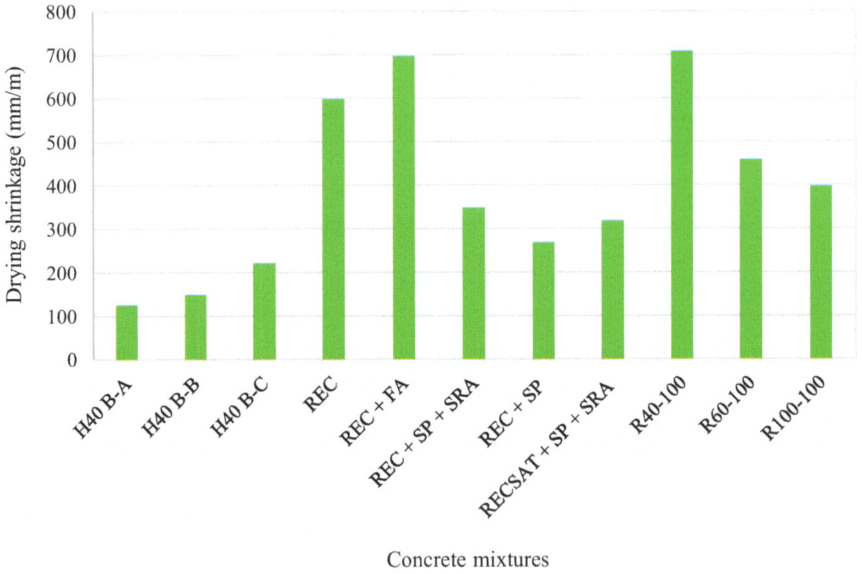

Fig. 7 Drying shrinkage of concrete mixtures containing recycled aggregates

3.3.2 Recycled Aggregates

The comparison of the concrete mixtures having recycled aggregates involved the selection of 11 distinct mixtures, as detailed in Table 8.

The comparative analysis of the drying shrinkage values among the selected concrete mixtures incorporating recycled aggregates reveals notable variations. Fig. 7 highlights that the H40 B mixtures exhibited the lowest drying shrinkage values. This can be owing to the replacement of coarse natural aggregates by recycled aggregates in varying proportions, while the sand component remained unaltered in these mixtures.

On the other hand, the REC and REC + FA mixtures displayed higher drying shrinkage values, as the entire natural aggregates were substituted with recycled aggregates. Similarly, the R40–100, R60–100, and R100–100 mixtures illustrated elevated drying shrinkage values due to the complete replacement of natural gravel by recycled aggregates. Additionally, it was observed that employing recycled aggregates sourced from concrete with higher strength levels led to reduced drying shrinkage in the concrete mixtures.

Among the studied mixtures, the REC + SP + SRA, REC + SP, and RECSAT + SP + SRA mixtures provided relatively low drying shrinkage values. This can be related to the replacement of all the natural sand and approximately 50% of gravel by recycled aggregates. In addition, the incorporation of shrinkage-reducing additives in two of the concrete mixtures contributed to the reduction in the drying shrinkage values.

3.3.3 Slag

The assessment of the drying shrinkage was also done for the concrete mixtures having slag. Four concrete mixtures were selected from the articles, addressing the drying shrinkage, and are presented in Table 9.

The findings shown in Fig. 8 depict that majority of the concrete mixtures exhibited low levels of the drying shrinkage, with the exception of the C25/30 mixture. The C25/30 mixture represents a conventional concrete composition where all natural aggregates were substituted with slag, thereby accounting for the elevated drying shrinkage observed. Additionally, this particular mixture utilized the CEM II type of cement, which may incorporate components such as slag, fly ash, limestone, and silica fume.

In contrast, low drying shrinkage values of the self compacting concrete mixtures (designated by SCC in Table 9 and Fig. 8) can be attributed to two primary

Table 9 Concrete mixtures containing slag

Article no.	Concrete mixture	Explanation
11	SCC-20	Self-compacting concrete mixtures containing fly ash, limestone filler, and 20%, 40%, and 60% slag
	SCC-40	
	SCC-60	
13	C25/30	Concrete mixture with strength class C25/30 having steel slag as aggregates

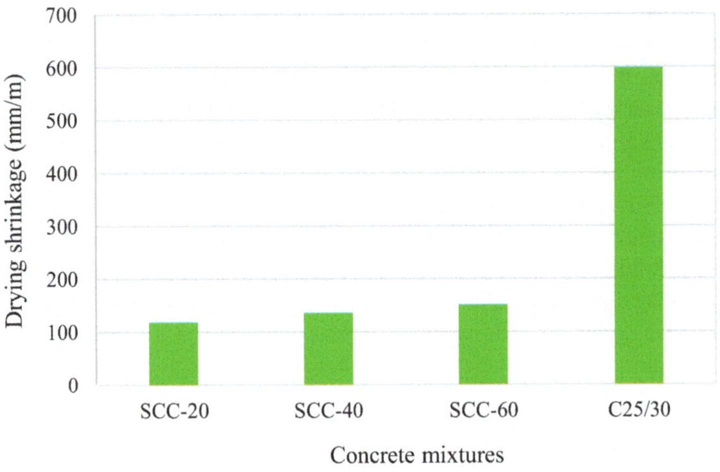

Fig. 8 Drying shrinkage of concrete mixtures containing slag

Table 10 Selected concrete mixtures for creep comparison

Article no.	Concrete mixture	Explanation
2	H40 B-A	Concrete mixtures containing 20%, 50%, and 100% recycled aggregates
	H40 B-B	
	H40 B-C	
8	G20	Concrete mixture where 20% of gravel was replaced by bottom ash
	G20 + S20	Concrete mixture where 20% of gravel and 20% of sand were replaced by bottom ash

factors. Firstly, the absence of recycled aggregates in the self-compacting concrete mixtures contributed to the mitigation of the drying shrinkage. Secondly, the self-compacting concrete mixtures featured lower proportions of slag, fly ash, and limestone in comparison to the C25/30 mixture, further contributing to their reduced drying shrinkage characteristics.

3.4 Creep

The investigation of the creep, as a concrete property, has received relatively less attention compared to other properties such as the compressive strength, resulting in a limited number of concrete mixtures selected for comparison. Table 10 summarizes five concrete mixtures, along with their designations, corresponding explanations, and the sources from which they were obtained.

Based on the comparison of the results, it can be concluded that the utilization of recycled aggregates had a detrimental impact on the creep behavior of the mixtures.

Increased replacement ratios of natural fine and coarse aggregates by recycled aggregates led to higher levels of the creep in the mixtures.

The comparative analysis in Fig. 9 shows a consistent pattern in the distribution of the creep values among the studied concrete mixtures, with the exception of the G20 mixture incorporating bottom ash, which demonstrated the lowest creep value. The observed increase in the creep values across the concrete mixtures correlates with the proportional increase in the recycled aggregates and bottom ash content.

Among the mixtures, the H40 B-C mixture displayed the highest creep value, attributed to the complete replacement of natural coarse aggregates by recycled aggregates. This substantial replacement proportion of coarse aggregates in the H40 mixtures, in comparison to the G20 and G20 + S20 mixtures, provides a plausible explanation for the elevated creep values.

3.5 *Modulus of Elasticity*

The modulus of elasticity stands out as a prominently evaluated property of concrete across the scientific articles reviewed in the literature. From the literature, 22 concrete mixtures were selected for the purpose of comparing the modulus of elasticity values. These mixtures were meticulously categorized and grouped based on the specific replacement material employed in their composition.

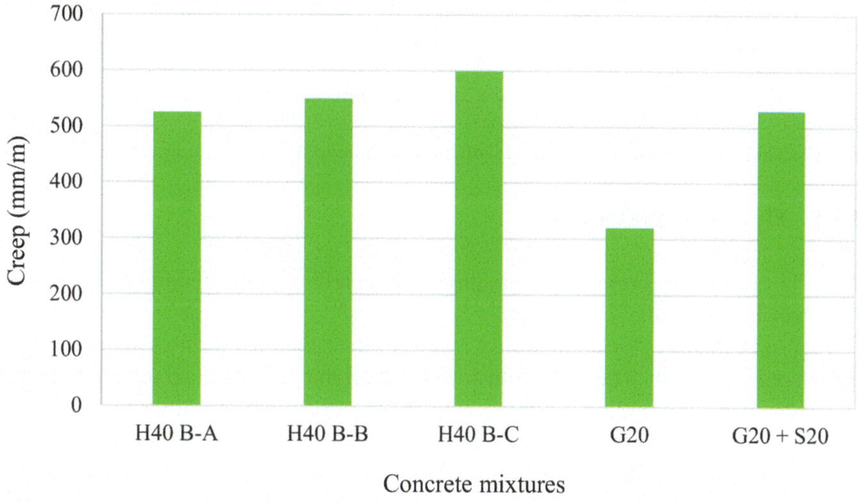

Fig. 9 Creep of concrete mixtures containing bottom ash or recycled aggregates

Table 11 Concrete mixtures containing fly ash or bottom ash

Article no.	Concrete mixture	Explanation
3	REC + FA	Concrete mixture containing 100% recycled aggregates and 38% fly ash
8	G20	Concrete mixture where 20% of gravel was replaced by bottom ash
	G20 + S20	Concrete mixture where 20% of gravel and 20% of sand were replaced by bottom ash
10	ECO-1	Concrete mixture containing limestone powder and 15% fly ash
	ECO-2	Concrete mixture containing limestone powder and 33% fly ash
	ECO-3	Concrete mixture containing limestone powder and 50% fly ash

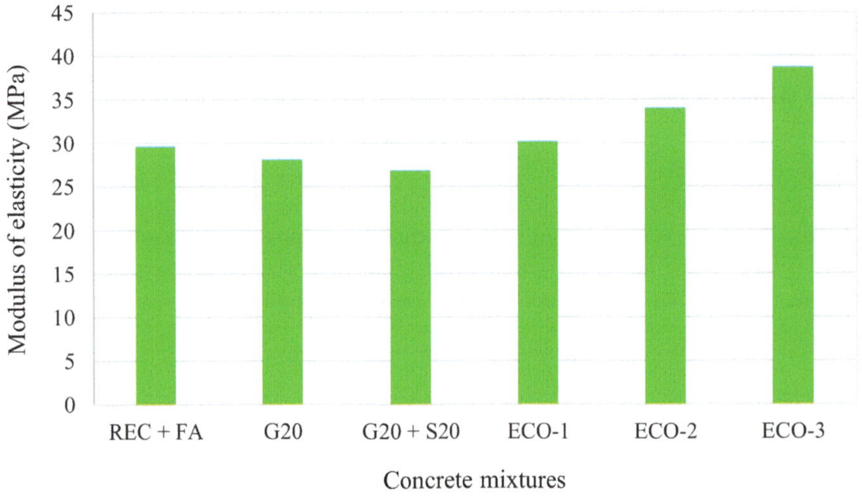

Fig. 10 Modulus of elasticity of concrete mixtures containing fly ash or bottom ash

3.5.1 Fly Ash or Bottom Ash

Table 11 presents information on six distinct concrete mixtures incorporating either fly ash or bottom ash.

Figure 10 displays the comparative results of the concrete mixtures incorporating fly ash or bottom ash. The figure illustrates that all the mixtures had the modulus of elasticity values exceeding 25 MPa, with the majority falling within the range of 25–30 MPa. The ECO (eco-friendly) mixtures indicated slightly higher values compared to the other mixtures. This discrepancy could be because of the utilization of significant quantities of limestone powder in conjunction with fly ash in the concrete mixtures. Moreover, the exclusion of recycled aggregates in the concrete mixture may contribute to the observed increase in the modulus of elasticity.

The lower modulus of elasticity values in certain concrete mixtures can be due to the replacement of natural aggregates by alternative materials, such as bottom ash or recycled aggregates. The REC + FA mixture, which had recycled aggregates, serves as an example. Likewise, the G20 and G20 + S20 mixtures replaced a portion of natural aggregates by bottom ash. Analyzing Fig. 10 reveals that an escalation in the replacement ratio for the G20 mixture correlated with a decrease in the modulus of elasticity of the mixture.

Table 12 outlines the concrete mixtures considered for the study on their modulus of elasticity containing recycled aggregates and/or ash. Fig. 11 depicts the comparison of the modulus of elasticity among the concrete mixtures having recycled aggregates.

The comparative analysis of the concrete mixtures points out that those mixtures incorporating recycled aggregates showed higher modulus of elasticity values in comparison to the mixtures containing fly ash or bottom ash. Specifically, the R40–100, R60–100, and R100–100 mixtures exhibited the highest modulus of elasticity values among the concrete mixtures, as shown in Fig. 11. This phenomenon could potentially be owing to the use of recycled aggregates derived from high-strength concrete. In contrast, the remaining concrete mixtures demonstrated lower values of the modulus of elasticity, which can be due to the incorporation of alternative replacement materials or an increased proportion of recycled aggregates. An example of this trend can be seen in the H-0.45ARS mixtures, where a gradual decrease in their modulus of elasticity occurred with an increase in the proportion of recycled aggregates. Similarly, in the case of the H40 mixtures, where natural coarse aggregates were partially replaced by recycled aggregates, a decrease in the modulus of elasticity was observed. The REC and REC + FA mixtures illustrated low values of the modulus of elasticity, likely because of the complete replacement of natural aggregates by recycled aggregates, as well as the high volume of fly ash in the REC + FA mixture. These adverse effects, however, are mitigated by the use of a larger amount of cement compared to the other concrete mixtures.

Table 12 Concrete mixtures containing recycled aggregates

Article no.	Concrete mixture	Explanation
2	H40 B-A	Concrete mixtures containing 20%, 50%, and 100% recycled aggregates
	H40 B-B	
	H40 B-C	
3	REC	Concrete mixture containing 100% recycled aggregates
	REC + FA	Concrete mixture containing 100% recycled aggregates and 38% fly ash
5	H-0.45ARS-A	Concrete mixtures with a w/c ratio of 0.45 and containing 20%, 50%, and 100% recycled aggregates
	H-0.45ARS-B	
	H-0.45ARS-C	
15	R40–100	Concrete mixtures containing 100% recycled aggregates with compressive strengths of 40, 60, and 100 MPa
	R60–100	
	R100–100	
19	C-100-100	Concrete mixture containing 100% recycled aggregates

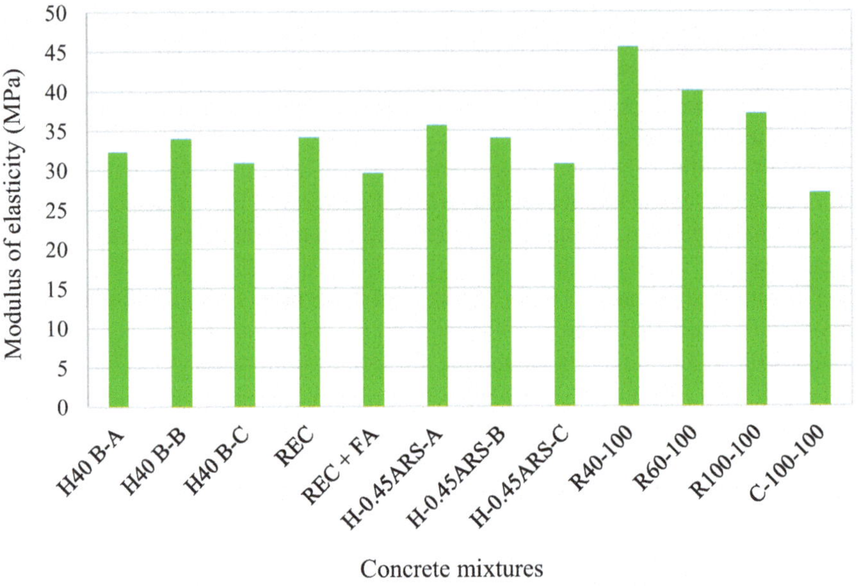

Fig. 11 Modulus of elasticity of concrete mixtures containing recycled aggregates

Table 13 Concrete mixtures containing slag and/or other additives

Article no.	Concrete mixture	Explanation
7	A-SF	Concrete mixtures containing silica dust or metakaolin
	A-MK	
11	SCC-20	Concrete mixtures containing fly ash, limestone filler and 20, 40, and 60% slag
	SCC-40	
	SCC-60	

3.5.2 Slag and/or Other Additives

Five specific mixtures were selected from the literature, as presented in Table 13. Figure 12 displays the comparison of the modulus of elasticity among various concrete mixtures incorporating slag and/or other additives.

The analysis revealed that the concrete mixtures having slag as a replacement material displayed higher modulus of elasticity values compared to the mixtures incorporating other additives such as silicate or metakaolin. One possible explanation for this observation is that in the A-SF and A-MK mixtures, 50% of natural aggregates was substituted with recycled aggregates. However, it should be noted that the influence of silicate powder or metakaolin on the concrete modulus of elasticity cannot be entirely disregarded.

The addition of fly ash along with high volumes of slag may account for the elevated modulus of elasticity values observed in the self-compacting concrete

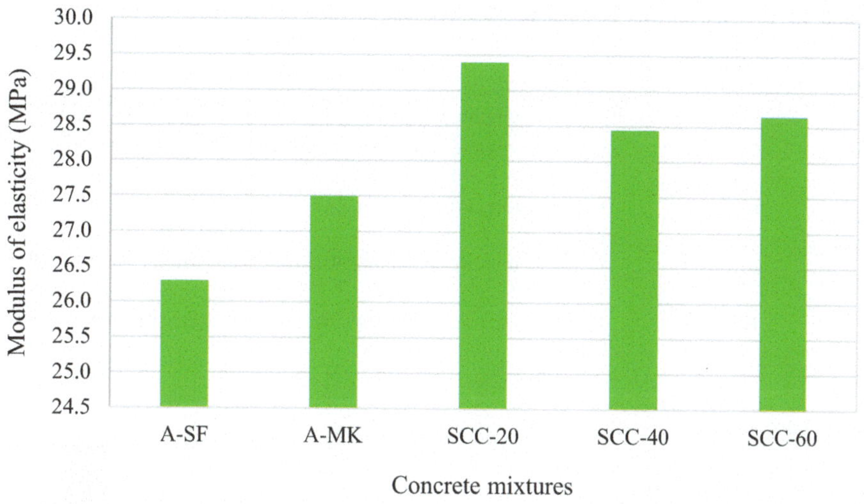

Fig. 12 Modulus of elasticity of concrete mixtures containing slag and/or other additives

mixtures (represented by SCC in the figure). These mixtures were formulated with the CEM II type of cement, which may contain fly ash, limestone, slag, and silicate powder, all of which could contribute to higher modulus of elasticity values in the concrete mixtures.

3.6 Carbonation Depth

In total, 12 concrete mixtures were identified for the purpose of carbonation depth comparison. These mixtures were categorized into two groups: one comprised of mixtures having recycled aggregates and the other one composed of mixtures containing either fly ash or bottom ash.

In general, the selected concrete mixtures displayed relatively shallow levels of the carbonation depth. However, an upward trend in the carbonation depth values was observed with an increase in the proportion of fly ash and recycled aggregates in the concrete mixtures. The use of CEM III type of cement in conjunction with recycled aggregates, fly ash, or other replacement materials was found to contribute to an elevated carbonation depth in the mixtures.

3.6.1 Fly Ash or Bottom Ash

The concrete mixtures containing fly ash or bottom ash considered for the study of the carbonation depth can be found in Table 14. Figure 13 illustrates the carbonation depths of the selected concrete mixtures after 28 days.

Table 14 Concrete mixtures containing fly ash or bottom ash

Article no.	Concrete mixture	Explanation
8	G20	Concrete mixture where 20% of gravel was replaced by bottom ash
	G20 + S20	Concrete mixture where 20% of gravel and 20% of sand were replaced by bottom ash
14	LR30-A	Concrete mixture containing 24% fly ash and 30% recycled aggregates
	MR30-A	Concrete mixture containing 30% recycled aggregates and 47% fly ash
	HR30-A	Concrete mixture containing 56% recycled aggregates and 47% fly ash

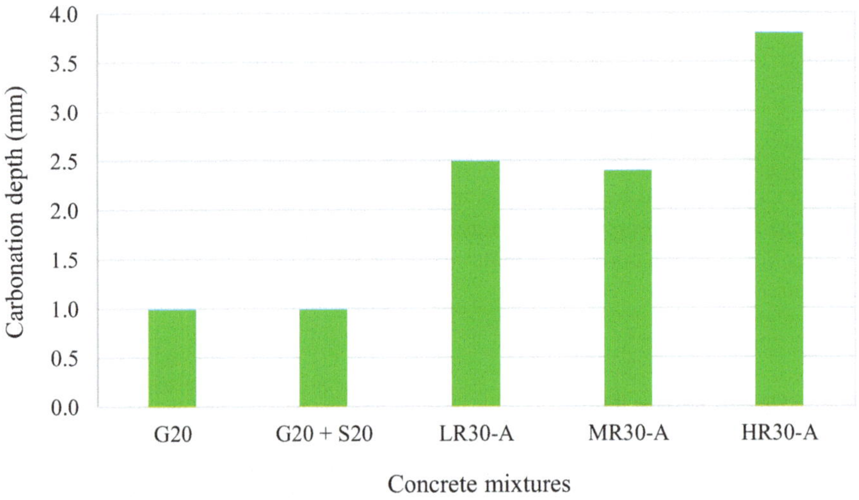

Fig. 13 Carbonation depth of concrete mixtures containing fly ash or bottom ash

All the investigated concrete mixtures indicated low carbonation depth values, with a gradual increase observed as the proportion of fly ash in the concrete mixture increased. Figure 13 shows various concrete mixtures incorporating different percentages of recycled aggregates, except for the mixtures G20 and G20 + S20, which utilized slag as aggregates.

In comparison to the other selected concrete mixtures, the mixtures G20 and G20 + S20 demonstrated the lowest carbonation depth values. This can be owing to the absence of recycled aggregates in these mixtures, unlike the LR30-A, MR30-A, and HR30-A mixtures. Additionally, the higher cement content in G20 and G20 + S20 compared to LR30-A, MR30-A, and HR30-A could also contribute to the lower carbonation depth values. Importantly, an increase in the carbonation depth was seen as the amount of fly ash increased and the cement content decreased in the LR30-A, MR30-A, and HR30-A mixtures.

3.6.2 Recycled Aggregates

The specific mixtures incorporating recycled aggregates, as outlined in Table 15, were selected for the examination of the carbonation depth.

Figure 14 depicts notable variations in the carbonation depth values among the concrete mixtures. In particular, the H-0.45ARS mixtures indicated the highest carbonation depth values compared to the other mixtures. This disparity in the carbonation depth values can be attributed to several factors, primarily linked to the alternative materials employed in the concrete mixtures. One possible explanation lies in the substantial amount of recycled aggregates incorporated into the H-0.45ARS mixtures. Moreover, the utilization of the CEM III type of cement, along with the complete replacement of natural aggregates by recycled aggregates in the C-100-100 mixture, may have contributed to an increased carbonation depth. On the other hand, the LR30-A, MR30-A, and HR30-A mixtures, which included some of these materials, showed lower carbonation depth values. Likewise, the C-100-100 mixture, using the CEM III type of cement containing slag, can also be considered in this hypothesis.

3.7 *Workability*

The assessment of the workability was addressed utilizing selected concrete mixtures from the articles in the literature, as presented in the following subsections. They were carefully selected, categorized, and subsequently divided into three distinct groups according to the replacement materials employed.

Table 15 Concrete mixtures containing recycled aggregates

Article no.	Concrete mixture	Explanation
5	H-0.45ARS-A	Concrete mixtures with a w/c ratio of 0.45 and containing 20, 50, and 100% recycled aggregates
	H-0.45ARS-B	
	H-0.45ARS-C	
14	LR30-A	Concrete mixtures containing 24% fly ash and 30% recycled aggregates
	MR30-A	Concrete mixture containing 30% recycled aggregates and 47% fly ash
	HR30-A	Concrete mixture containing 56% recycled aggregates and 47% fly ash
19	C-100-100	Concrete mixture containing 100% recycled aggregates

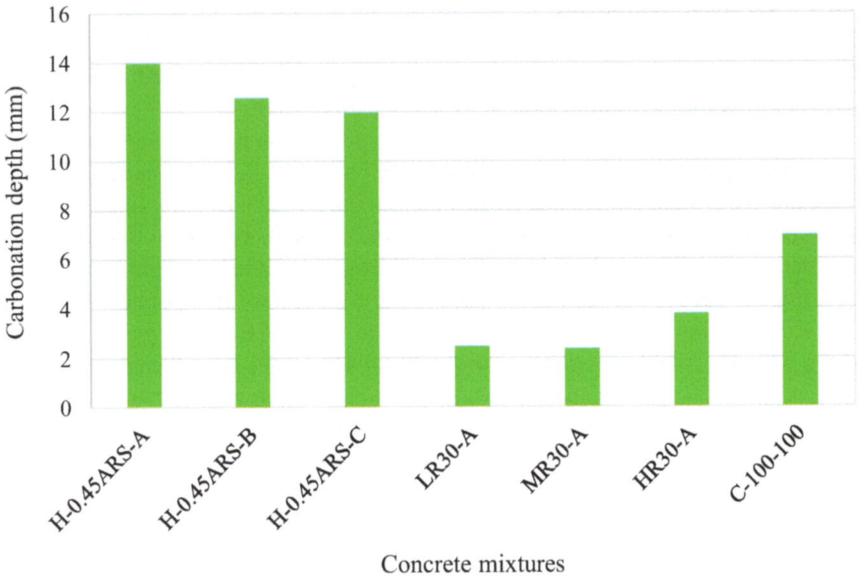

Fig. 14 Carbonation depth of concrete mixtures containing recycled aggregates

Table 16 Concrete mixtures containing fly ash or bottom ash

Article no.	Concrete mixture	Explanation
3	REC + FA	Concrete mixture where 20% of gravel was replaced by bottom ash
8	G20	Concrete mixture where 20% of gravel was replaced by bottom ash
	G20 + S20	Concrete mixture where 20% of gravel and 20% sand were replaced by bottom ash
9	MN	Concrete mixture containing 47% fly ash
	MR30	Concrete mixture containing 30% recycled aggregates and 47% fly ash
	HN	Concrete mixture containing 56% fly ash
12	G-FA	Concrete mixture containing 25% fly ash
17	RAC_FA	Concrete mixture containing fly ash and 100% recycled aggregates
	NAC_FA	Concrete mixture containing fly ash and 100% natural aggregates
	NAC_AAFA	Concrete mixture containing alkaline-activated fly ash and 100% natural aggregates

3.7.1 Fly Ash or Bottom Ash

The study involved concrete mixtures containing either fly ash or bottom ash, with a total of 10 concrete mixtures being selected for comparison. The details of these mixtures are provided in Table 16. The table serves as a comprehensive overview of the various concrete mixtures under consideration.

The findings pertaining to the concrete mixtures incorporating fly ash are demonstrated in Fig. 15. As shown in the figure, majority of the concrete mixtures illustrated values within the range of 100 mm and 210 mm, aligning with the slump test classes S3 and S4.

The utilization of fly ash as a substitute material in the concrete mixtures is associated with reduced slump test values. The introduction of fly ash into the concrete mixtures has yielded an increment in the slump test values, with a proportional relationship between the quantity of fly ash employed and the resulting slump test values. The MN and HN mixtures serve as examples supporting this observation.

The MN, MR30, and HN mixtures displayed the slump test values corresponding to class S5, surpassing those of the remaining mixtures. A plausible explanation for these elevated slump test values lies in the increased fly ash content and the relatively increased w/c ratio implemented in these concrete mixtures. In contrast, the remaining mixtures indicated lower slump test values due to either diminished fly ash content or reduced w/c ratios in comparison to the MN, MR30, and HN mixtures.

The incorporation of alternative cementitious materials into concrete mixtures can have an impact on the workability of the resulting mixtures. For instance, in the case of the G20 and G20 + S20 mixtures, the slump test values were found to be the lowest. This decrease in the workability can be attributed to the use of relatively small proportions of bottom ash as a substitute in these mixtures.

Similarly, the RAC_FA and NAC_FA mixtures depicted the slump test values comparable to those of the G20 and G20 + S20 mixtures. However, the RAC_FA and NAC_FA mixtures had a high content of fly ash and were characterized by low

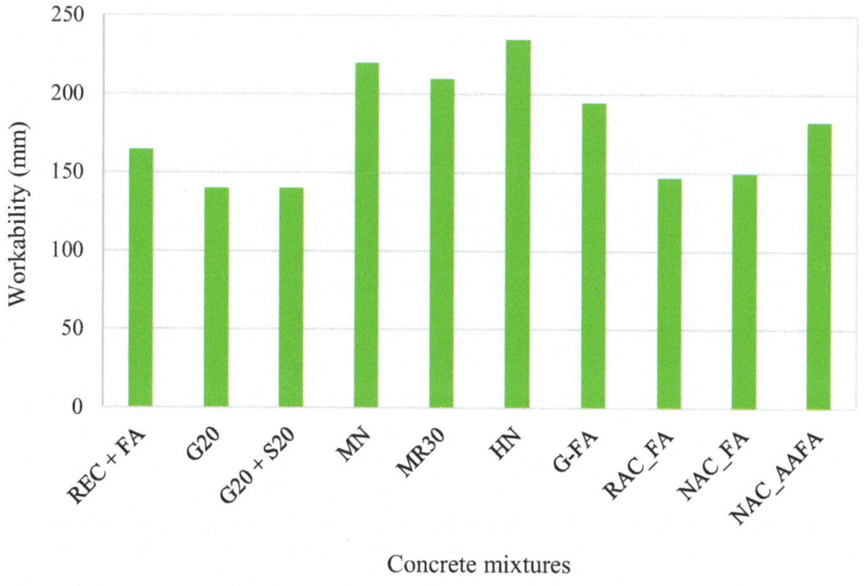

Fig. 15 Workability of concrete mixtures containing fly ash or bottom ash

w/c ratios. The diminished workability witnessed in the RAC_FA and NAC_FA mixtures, despite the substantial presence of fly ash, could potentially be linked to the usage of the CEM II type of cement in their production.

Additionally, the G-FA mixture exhibited a slump test value within the S4 classification. This outcome could be because of the elevated cement content and the incorporation of fly ash in the mixture.

In summary, the inclusion of alternative cementitious materials, such as fly ash or bottom ash, in the concrete mixtures can influence the workability of the resulting mixture, as evidenced by the variation in the slump test values observed in the mixtures. The specific characteristics of the cement type employed in the mixture may also contribute to the resulting effects on the workability.

3.7.2 Recycled Aggregates

Table 17 presents various concrete mixtures having recycled aggregates utilized in this study to evaluate their workability, while their comparisons are illustrated in Fig. 16.

The findings revealed a remarkable pattern akin to the composition of the concrete mixtures incorporating fly ash. Majority of the concrete mixtures indicated the slump test values below the thresholds denoted by the slump test classes S3 and S4.

Table 17 Concrete mixtures containing recycled aggregates

Article no.	Concrete mixture	Explanation
2	H40 B-A	Concrete mixtures containing 20, 50, and 100% recycled aggregates
	H40 B-B	
	H40 B-C	
3	REC	Concrete mixture containing 100% recycled aggregates
	REC + FA	Concrete mixture containing 100% recycled aggregates and 38% fly ash
4	REC + SP + SRA	Concrete mixture containing 100% recycled aggregates, superplasticizer, and shrinkage-reducing admixture
	REC + SP	Concrete mixture containing 100% recycled aggregates and superplasticizer
	RECSAT + SP + SRA	Concrete mixture containing 100% saturated recycled aggregates, superplasticizer, and shrinkage-reducing admixture
9	MR30	Concrete mixture containing 30% recycled aggregates and 47% fly ash
12	REC-A	Concrete mixture containing 100% recycled aggregates
	G-REC-SL	Concrete mixture containing 100% recycled aggregates and slag
17	RAC_FA	Concrete mixture containing 100% recycled aggregates and fly ash
19	C-100-100	Concrete mixture containing 100% recycled aggregates

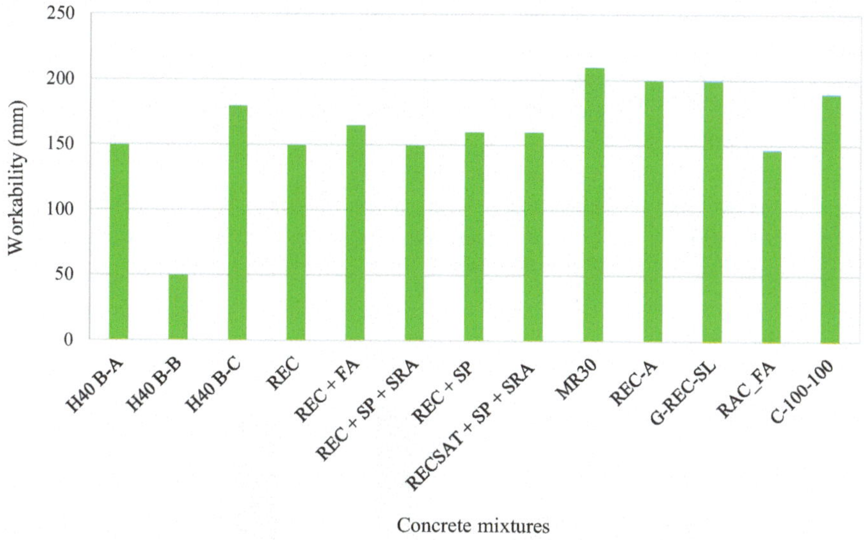

Fig. 16 Workability of concrete mixtures containing recycled aggregates

Particularly, the H40 B-B mixture showed the lowest slump test value, measuring 50 mm, and fell within the slump test class S2.

In contrast, the MR30, REC-A, and G-REC-SL mixtures exhibited the highest slump test values among all the studied concrete mixtures. This disparity can be owing to the inclusion of approximately 30% recycled aggregates in these three mixtures, unlike the remaining mixtures which contained higher proportions of recycled aggregates, consequently resulting in lower slump test values. However, it is noteworthy that the C-100-100 mixture displayed a relatively high slump test value despite the complete replacement of natural aggregates by recycled aggregates. One plausible explanation for this discrepancy could be the utilization of the CEM III type of cement, which incorporates slag.

The use of recycled aggregates in conjunction with additives, such as superplasticizers and shrinkage-reducing agents, resulted in lower slump test values for the concrete mixtures. Furthermore, the incorporation of fly ash along with recycled aggregates also contributed to slightly diminished slump test values and compromised workability of the concrete mixtures. This observation is consistent with prior research on the subject, thereby pointing out that the inclusion of fly ash tends to impede the workability of the concrete mixtures.

3.7.3 Slag and/or Other Additives

Table 18 provides a comprehensive overview of five distinct concrete mixtures, each having different types of replacement materials. The comparative analysis of the workability of these concrete mixtures is depicted in Fig. 17.

Table 18 Concrete mixtures containing slag and/or other additives

Article no.	Concrete mixture	Explanation
7	A-SF	Concrete mixtures containing silica dust or metakaolin
	A-MK	
12	G-SL	Concrete mixture containing slag
	G-REC-SL	Concrete mixture containing 100% recycled aggregates and slag
13	C25/30	Concrete mixture with strength class C25/30 and containing steel slag as aggregates

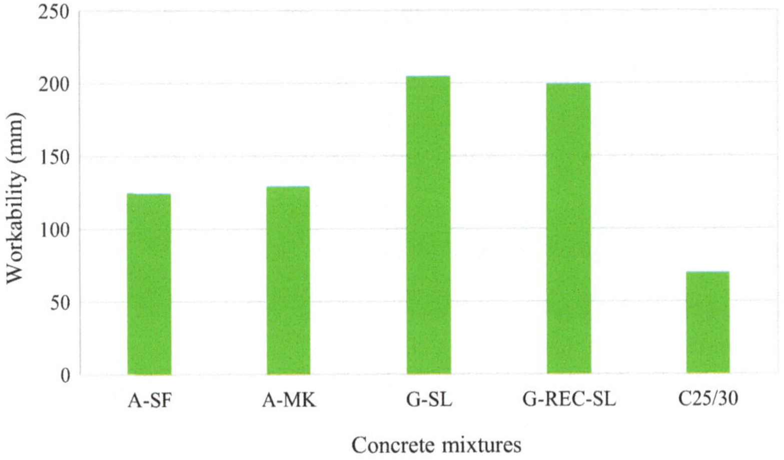

Fig. 17 Workability of concrete mixtures containing slag and/or other additives

The G-SL and G-REC-SL mixtures illustrated the most significant slump test values when compared to the remaining concrete mixtures. The enhanced slump test values can be because of the utilization of slag in relatively smaller proportions. Alternatively, the addition of recycled aggregates resulted in a slight reduction in the slump test value. However, it should be noted that an increase in the slag content at higher volumes led to an unfavorable impact on the workability of the concrete mixtures, as exemplified by the C25/30 mixture where slag served as the exclusive aggregates.

In the cases of the A-SF and A-MK mixtures, the slump test values were witnessed to fall below the S3 classification. This occurrence can be owing to the presence of limited quantities of silica fume or metakaolin along with substantial amounts of recycled aggregates.

4 Conclusions

A comprehensive comparison was carried out in this study to investigate the impact of alternative replacement materials on the properties and performance of different concrete mixtures. The results of the comparison revealed that the incorporation of fly ash, blast furnace slag, or bottom ash in the concrete mixtures had both positive and negative effects. Fly ash and bottom ash proved to be suitable substitutes for cement, effectively reducing carbon dioxide emissions. Additionally, these materials exhibited the potential to enhance the concrete compressive strength, tensile strength, and modulus of elasticity. However, it is crucial to limit the proportion of these materials to prevent any adverse effects on other concrete properties.

The utilization of elevated proportions of fly ash or bottom ash in the concrete mixtures was found to adversely affect the modulus of elasticity and workability of the concrete mixtures. Moreover, higher volumes of fly ash or bottom ash contributed to an increase in the creep, carbonation depth, and drying shrinkage values in the concrete mixtures. Similar effects on the concrete mixtures were observed when incorporating slag, akin to the usage of fly ash or bottom ash. The concrete mixtures containing slag demonstrated higher values of the compressive strength and modulus of elasticity. However, as the proportion of the slag replacement increased, detrimental effects were observed. In particular, an escalation in the slag content led to poorer performance in terms of the creep and drying shrinkage of the concrete mixtures.

The substitution of natural aggregates with recycled aggregates generally had an adverse impact on the properties and performance of concrete. In majority of the concrete mixtures containing varying percentages of recycled aggregates, lower values of the modulus of elasticity and strength were witnessed compared to those incorporating natural aggregates. In addition, the use of high quantities of recycled aggregates led to increased creep and drying shrinkage in the concrete mixtures. However, it should be mentioned that the workability of the concrete mixtures was not affected to the same extent as other properties of the concrete mixtures. While the use of recycled aggregates offers environmental benefits, it is crucial to adhere to acceptable limits to avoid compromising the structural properties of concrete. Considering environmentally friendly alternative materials such as crushed rock may be a viable approach to achieve the desired results. Furthermore, incorporating other replacement materials that enhance concrete properties and mitigate the adverse effects associated with the addition of recycled aggregates can provide a potential solution to address these challenges.

The comprehensive analysis of the results leads to the conclusion that the selection of replacement materials in the concrete mixtures significantly influences the properties and structural performance of climate-improved concrete. When a small proportion of recycled aggregates replaced natural aggregates, the impact on the properties of the concrete mixtures was found to be relatively minor. However, as the volume of recycled aggregates increased, a noticeable decline in the performance of the concrete mixtures was observed. While the utilization of recycled aggregates in climate-improved concrete offers an effective means to reduce the carbon footprint, it is crucial to employ a combination of complementary materials, such as slag or fly ash, to mitigate any potential adverse effects and optimize the overall performance of the concrete mixtures.

References

1. Swedish Concrete, *Concrete for Sustainable Urban Development.* (in Swedish) (2022a) Retrieved 22 May, 2022, https://www.svenskbetong.se/om-betong
2. P.G. Burström, K. Nilvér, *Building Materials: Production, Properties, and Use*, 3rd edn. (Student Literature, Lund, 2018)
3. Thomas Betong. *Concrete and Carbon Dioxide.* Retrieved 22 May, 2022., https://thomasbetong.se/om-betong/
4. M.A.O. Mydin, P. Jagadesh, A. Bahrami, S.S. Majeed, A. Dulaimi, R. Omar, Study on fresh and hardened state properties of eco-friendly foamed concrete incorporating waste soda-lime glass. Sci. Rep. **14**, 18733 (2024). https://doi.org/10.1038/s41598-024-69572-4
5. Swedish National Board of Housing, Building, and Planning (Boverket). *Greenhouse Gas Emissions from the Building and Real Estate Sector,* (in Swedish) (2021), Retrieved 22 May 2022, https://www.boverket.se/sv/byggande/hallbart-byggande-och-forvaltning/miljoindikatorer%2D%2D-aktuell-status/vaxthusgaser/
6. Swedish Concrete, *Carbon Dioxide Emissions.* (in Swedish) (2022b) Retrieved 22 May, 2022, https://www.svenskbetong.se/om-betong/prefab/miljo-och-hallbarhet/koldioxidutslapp
7. Swedish Environmental Protection Agency, *Sweden's Climate Goals and Climate Policy Framework.* (in Swedish), Retrieved 22 May, 2022., https://www.naturvardsverket.se/amnesomraden/klimatomstallningen/sveriges-klimatarbete/sveriges-klimatmal-och-klimatpolitiska-ramverk
8. Swedish Concrete, *Guidance on Climate-Improved Concrete. (in Swedish)* (Swedish Concrete, Stockholm, 2022c) https://www.svenskbetong.se/component/edocman/?task=document.viewdoc&id=60&Itemid=
9. A.I. Celik, U. Tunc, A. Bahrami, M. Karalar, M.A.O. Mydin, T. Alomayri, Y.O. Özkılıc, Use of waste glass powder toward more sustainable geopolymer concrete. J. Mater. Res. Technol. **24**, 8533–8546 (2023). https://doi.org/10.1016/j.jmrt.2023.05.094
10. C. Aksoylu, Y.O. Özkılıç, A. Bahrami, S.A. Yıldızel, I.Y. Hakeem, N. Özdöner, B. Basaran, M. Karalar, Application of waste ceramic powder as a cement replacement in reinforced concrete beams toward sustainable usage in construction. Case Stud. Constr. Mater. **19**, e02444 (2023). https://doi.org/10.1016/j.cscm.2023.e02444
11. M. Wagh, U. Waghe, A. Bahrami, K. Ansari, Y.O. Özkılıç, A. Nikhade, Experimental investigation of mechanical and durability performances of self-compacting concrete blended with bagasse ash, metakaolin, and glass fiber. Front. Mater. **11**, 1351554 (2024). https://doi.org/10.3389/fmats.2024.1351554
12. I. Yüksel, T. Bilir, Ö. Özkan, Durability of concrete incorporating non-ground blast furnace slag and bottom ash as fine aggregate. Build. Environ. **42**(7), 2651–2659 (2007). https://doi.org/10.1016/j.buildenv.2006.07.003
13. A. Domingo-Cabo, C. Lázaro, F. López-Gayarre, M.A. Serrano-López, P. Serna, J.O. Castaño-Tabares, Creep and shrinkage of recycled aggregate concrete. Constr. Build. Mater. **23**(7), 2545–2553 (2009). https://doi.org/10.1016/j.conbuildmat.2009.02.018
14. V. Corinaldesi, G. Moriconi, Influence of mineral additions on the performance of 100% recycled aggregate concrete. Constr. Build. Mater. **23**(8), 2869–2876 (2009). https://doi.org/10.1016/J.CONBUILDMAT.2009.02.004

15. V. Corinaldesi, G. Moriconi, Recycling of rubble from building demolition for low-shrinkage concretes. Waste Manag. **30**(4), 655–659 (2010). https://doi.org/10.1016/J.WASMAN.2009.11.026
16. C. Thomas, J. Setién, J.A. Polanco, P. Alaejos, M. Sánchez De Juan, Durability of recycled aggregate concrete. Constr. Build. Mater. **40**, 1054–1065 (2012). https://doi.org/10.1016/j.conbuildmat.2012.11.106
17. M. Alwaeli, Application of granulated lead-zinc slag in concrete as an opportunity to save natural resources. Radiat. Phys. Chem. **83**, 54–60 (2013). https://doi.org/10.1016/j.radphyschem.2012.09.024
18. V. Radonjanin, M. Malešev, S. Marinković, A.E.S. al Malty, Green recycled aggregate concrete. Constr. Build. Mater. **47**, 1503–1511 (2013). https://doi.org/10.1016/j.conbuildmat.2013.06.076
19. G. Van Dr Wegen, U. Hofstra, J. Speerstra, Upgraded MSWI bottom ash as aggregate in concrete. Waste Biomass Valori. **4**(4), 737–743 (2013). https://doi.org/10.1007/s12649-013-9255-6
20. C. Lima, A. Caggiano, C. Faella, E. Martinelli, M. Pepe, R. Realfonzo, Physical properties and mechanical behaviour of concrete made with recycled aggregates and fly ash. Constr. Build. Mater. **47**, 547–559 (2013). https://doi.org/10.1016/J.CONBUILDMAT.2013.04.051
21. T. Proske, S. Hainer, M. Rezvani, C.A. Graubner, Eco-friendly concretes with reduced water and cement content—Mix design principles and application in practice. Constr. Build. Mater. **67**(PART C), 413–421 (2014). https://doi.org/10.1016/J.CONBUILDMAT.2013.12.066
22. M. Valcuende, F. Benito, C. Parra, I. Miñano, Shrinkage of self-compacting concrete made with blast furnace slag as fine aggregate. Constr. Build. Mater. **76**, 1–9 (2015). https://doi.org/10.1016/j.conbuildmat.2014.11.029
23. J. Turk, Z. Cotič, A. Mladenovič, A. Šajna, Environmental evaluation of green concretes versus conventional concrete by means of LCA. Waste Manag. **45**, 194–205 (2015). https://doi.org/10.1016/J.WASMAN.2015.06.035
24. L. Rondi, G. Bregoli, S. Sorlini, L. Cominoli, C. Collivignarelli, G. Plizzari, Concrete with EAF steel slag as aggregate: A comprehensive technical and environmental characterisation. Compos. Part B **90**, 195–202 (2016). https://doi.org/10.1016/j.compositesb.2015.12.022
25. C. Faella, C. Lima, E. Martinelli, M. Pepe, R. Realfonzo, Mechanical and durability performance of sustainable structural concretes: An experimental study. Cem. Concr. Compos. **71**, 85–96 (2016). https://doi.org/10.1016/J.CEMCONCOMP.2016.05.009
26. A. Gonzalez-Corominas, M. Etxeberria, Effects of using recycled concrete aggregates on the shrinkage of high performance concrete. Constr. Build. Mater. **115**, 32–41 (2016). https://doi.org/10.1016/j.conbuildmat.2016.04.031
27. G.L. Golewski, Improvement of fracture toughness of green concrete as a result of addition of coal fly ash. Characterization of fly ash microstructure. Mater. Charact. **134**, 335–346 (2017). https://doi.org/10.1016/J.MATCHAR.2017.11.008
28. S. Marinković, J. Dragaš, I. Ignjatović, N. Tošić, Environmental assessment of green concretes for structural use. J. Clean. Prod. **154**, 633–649 (2017). https://doi.org/10.1016/J.JCLEPRO.2017.04.015
29. M. Velay-Lizancos, I. Martinez-Lage, M. Azenha, J. Granja, P. Vazquez-Burgo, Concrete with fine and coarse recycled aggregates: E-modulus evolution, compressive strength and non-destructive testing at early ages. Constr. Build. Mater. **193**, 323–331 (2018). https://doi.org/10.1016/J.CONBUILDMAT.2018.10.209
30. A.T. Gebremariam, A. Vahidi, F. di Maio, J. Moreno-Juez, I. Vegas-Ramiro, A. Łagosz, R. Mróz, P. Rem, Comprehensive study on the most sustainable concrete design made of recycled concrete, glass and mineral wool from C&D wastes. Constr. Build. Mater. **273**, 121697 (2021). https://doi.org/10.1016/J.CONBUILDMAT.2020.121697

Optimization of Reinforced Concrete Structure of Multi-Story Residential Building in Sweden

Alireza Bahrami, Ali Mohamed Ahmed, and Obaidullah Rashid

1 Introduction

As the global population continues to grow, the demand for new residential buildings is significantly increasing. At the same time, there has been a rising need for housing and construction worldwide over the past few decades, leading to an increased demand for building materials. Concrete is the most important building material for multi-story buildings because of its high compressive strength, but it largely contributes to greenhouse gas emissions. To reduce carbon footprint and promote sustainability, it is vital to minimize material usage and adopt modern construction methods. Building structure plays a crucial role in ensuring safety and long-term sustainability within constructions. Traditionally, load-bearing walls are the conventional means of supporting roofs and floors. However, advancements in technology have introduced opportunities to enhance structural efficiency by replacing load-bearing walls with columns and beams. This method offers many advantages, including increased flexibility in design and construction [1, 2].

Replacing exterior and interior load-bearing walls with columns and beams can enhance both the appearance and functionality of a building, enabling a more open and spacious floor plan, facilitating larger rooms with fewer obstacles. Moreover, increased natural light and improved flow contribute to a brighter and more vibrant atmosphere. Additionally, columns and beams occupy less space compared to load-bearing walls, thus maximizing the available space within the same area [3].

To address the significant negative impact of concrete production on the environment due to its carbon dioxide emissions, construction methods are being improved. One approach involves determining optimal dimensions for load-bearing elements

A. Bahrami (✉) · A. M. Ahmed · O. Rashid
Department of Building Engineering, Energy Systems and Sustainability Science, Faculty of Engineering and Sustainable Development, University of Gävle, Gävle, Sweden
e-mail: alireza.bahrami@hig.se

A. Bahrami (ed.), *Advances in Sustainable Concrete for Construction*,
https://doi.org/10.1007/978-3-031-85052-3_2

like columns and beams based on the building's needs and strength requirements. By reducing these dimensions, less material is required, cutting down on construction energy usage, costs, and carbon dioxide emissions [4]. Utilizing this approach enables a reduction in material usage, offering advantages from both environmental and cost perspectives [5]. Additionally, employing columns and beams enhances design flexibility, facilitating the creation of expansive open spaces without the risk of structural failure.

Beams and columns are two of the most important building elements in a structure. These elements can combine both bending and compression to carry loads and provide stability to the entire structure. By understanding how loads affect beams and columns, how loads are distributed across the elements, and what type of bending and compression occurs as a result of these loads, engineers and builders can design beams and columns that are strong and efficient enough to bear the specific loads required [6].

Replacing load-bearing walls with columns and beams results in a structurally optimized building with lighter and more efficient load-bearing elements. These actions enhance the building's flexibility and the usability of its space [7].

Multi-story buildings show reinforced concrete (RC) floors outperforming cross-laminated timber in vertical deflection and utilization [8, 9]. Composite steel–concrete beams are commonly employed in construction for their superior strength, rigidity, and capacity to resist bending and shearing forces [10]. Prefabricated systems save time, boosting construction efficiency and reducing environmental impact [11]. This type of construction brings efficiency, less waste, and better safety [12].

The enhanced colliding bodies optimization algorithm has been used to minimize both construction costs and carbon dioxide emissions in RC frames [13]. Many of the existing optimized methods are specialized in the field of structural design, which consider all parameters as deterministic (deterministic-based structure optimization) [14].

Kulkarni and Bhusare [15] investigated the structural optimization of RC structures, with a focus on achieving cost-effective, robust, and safe designs while considering carbon dioxide emissions. Esfandiary et al. [16] introduced a new method for optimizing RC frames that can contribute to a more efficient and effective optimization of structural engineering. Leps [17] did a detailed review of how multi-objective optimization can enhance the design process for RC frames, providing insights into both theory and practical implementation. Kusuma et al. [18] examined concrete mixtures and reported how improved concrete quality can play an important role in developing concrete products with reduced environmental impact and contribute to sustainable development. de Medeiros and Kripka [19] assessed the optimization of RC columns according to different environmental impact assessment parameters. Ženíšek et al. [20] evaluated the optimization of RC structures in terms of cost and environmental impact.

The current chapter investigates the optimization of RC structure of a multi-story residential building by replacing its load-bearing walls with load-bearing columns and beams. A real building as reference building (building 1) located in Sandviken, Sweden, was adopted, and the StruSoft FEM-Design software was employed. First, modeling, designing, and analyzing building 1, which has load-bearing RC walls, were conducted. Then, the load-bearing RC walls of building 1 were replaced with load-bearing RC columns and beams (building 2). Stresses, deflections, utilization ratios, reaction forces, weights, and stability of both buildings were examined.

2 Methodology

The methodology of this study comprises modeling, analysis, and design of building 1 located in Sandviken, Sweden, and building 2 using the StruSoft FEM-Design software.

2.1 Building 1

In this study, building 1 is the reference building, which has been utilized to model a real project workflow. The building is a multi-story structure with eight stories and no basement. Stories two to eight are identical and have the same layout and design. However, the ground story has a different geometry compared to the other stories. Based on the technical drawings, the height of the building is 27.17 m, and the width of the ground story is 20.665 m. The seven upper stories are identical in both size and design. The building has a standard story height of 3 m from the top of the story slab to the bottom of the story slab on all stories. The building has a mono-pitched roof constructed with a double shed roof structure, meaning that it has two roofs meeting in the middle and forming an angle.

The structural framework of building 1 was built using RC, a widely utilized and robust building material. The load-bearing elements in the building are walls, columns, and story slabs, which together provide stability and load-bearing capacity in the building. The dimensions of these structural elements are as follows:

- The walls of the building have a thickness of 200 mm.
- The columns in the building are square-shaped 200 × 200 mm.
- The story slab in the building has a thickness of 250 mm.

Figure 1 illustrates the different views of building 1.

2.1.1 Drawings of Building 1

Figures 2 and 3 display the architectural plans, and Figs. 4 and 5 depict the load-bearing elements of the building. In this research, building 1 constructed with load-bearing RC walls was utilized as a basis. It was then transformed into a new model by replacing the load-bearing RC walls with load-bearing RC columns and beams. This process involved two main steps as presented below.

Northeast facade Northwest facade

Southeast facade Southwest facade

Fig. 1 Different views of building 1

2.2 *Modeling, Analysis, and Design of Building 1*

A three-dimensional model of building 1 (existing building) was created using the StruSoft FEM-Design software for the purpose of examining its stresses, deflections, utilization ratios, reaction forces, weights, and stability. The modeling process also incorporated relevant applied loads. In the process of constructing a building, it is essential to consider all the different loads it may experience over its operational lifetime. An important factor affecting the building's stability is its own weight, which is the weight the building has due to its materials and construction.

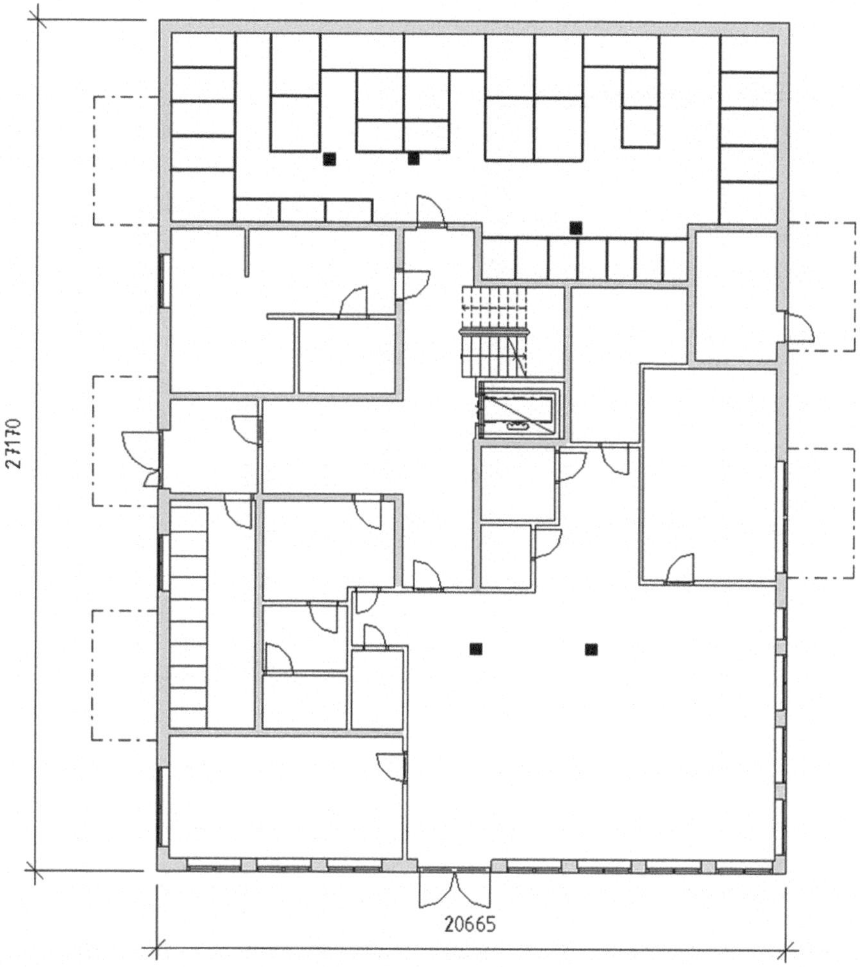

Fig. 2 Architectural plan of ground story (units: mm)

However, it is not only the self-weight that matters, but also other types of loads that can affect the building, such as live loads (temporary loads due to occupancy, furniture, snow, wind, etc.), must also be considered within the structure.

Table 1 summarizes the loads applied to the building. The StruSoft FEM-Design software generated the wind load, which was then applied to the structure. The software automatically combined the loads to determine the most unfavorable combination [2, 9].

After modeling building 1, the next step was to design its various elements. This step involved analyzing the structural components in each story and considering the utilization ratio of each individual load-bearing element. It is important to properly design various building elements to ensure the stability and durability of the

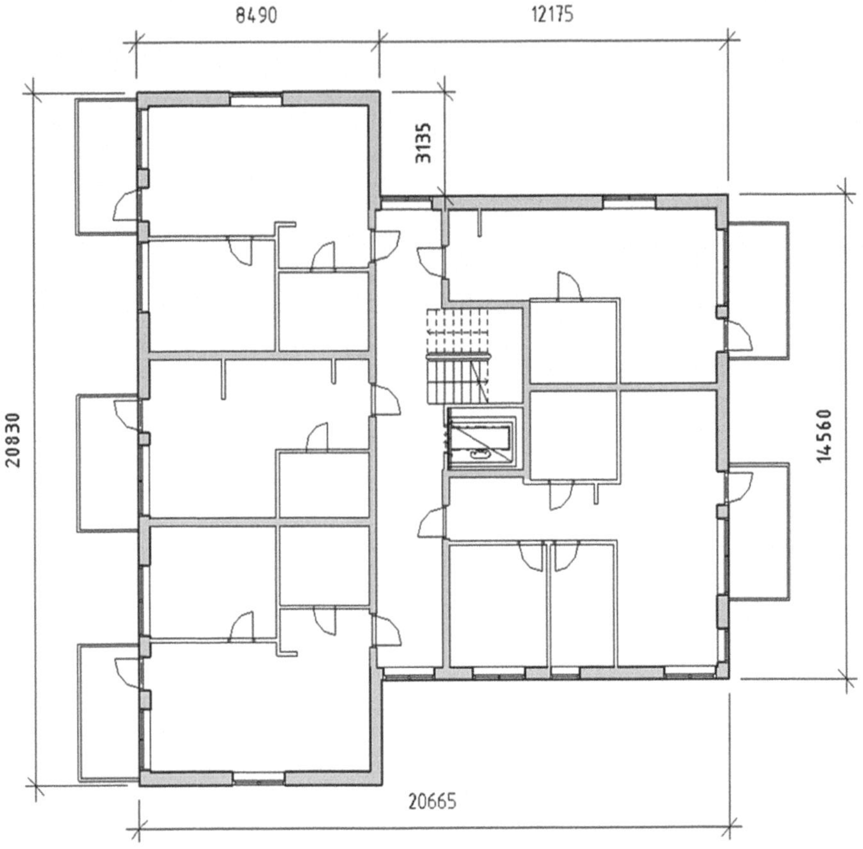

Fig. 3 Architectural plan of stories 2–8 (units: mm)

structure. For RC structures, it is necessary to design reinforcements to attain the desired structural performance. The size, spacing, and placement of reinforcements were adjusted until the structure achieved acceptable utilization ratios.

Once all the construction elements of building 1 were modeled and designed properly, the building took the shape as shown in Fig. 6.

2.3 Modeling, Analysis, and Design of Building 2

To evaluate the performance of the building with load-bearing RC columns and beams (building 2) instead of load-bearing RC walls (building 1), a replica of building 1 was constructed using the StruSoft FEM-Design software. By replacing the load-bearing walls with columns and beams, a new model of the structure was created. This modified model (building 2) was then subjected to identical loads and

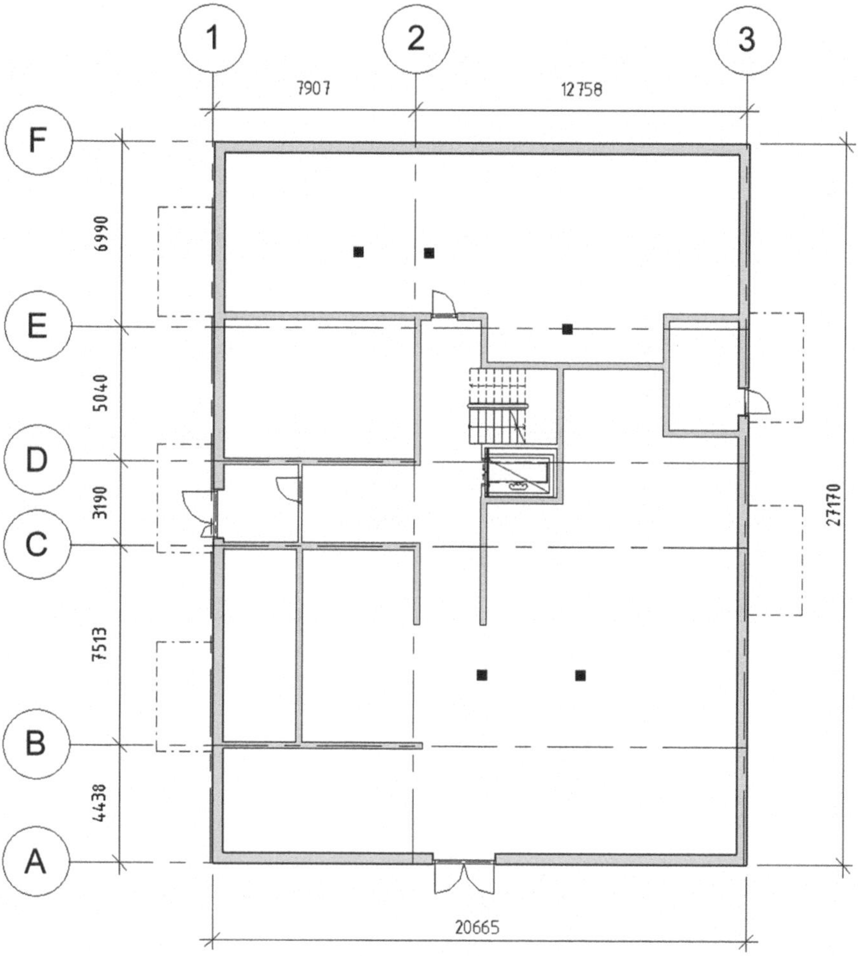

Fig. 4 Load-bearing elements of ground story (units: mm)

conditions as those applied to building 1. To model building 2, it was necessary to replace the load-bearing RC walls on all stories of building 1 with load-bearing RC columns and beams. The load-bearing RC walls on every story of building 1 were replaced with RC columns, each having cross-sectional dimensions of 300 × 300 mm. RC beams, with cross-sectional dimensions of 200 × 300 mm and 300 × 400 mm, were utilized to create a suitable structure for each story. The RC floor slabs, with a standard thickness of 250 mm, were also employed in this model. These slabs would support all floors of the building, ensuring a stable floor for each level. The specifications of the used RC elements are listed in Table 2 as well. The concrete strength class considered for all the elements was C30/37.

After modeling, analysis, and design of all the building's load-bearing elements, the building look liked as illustrated in Fig. 7.

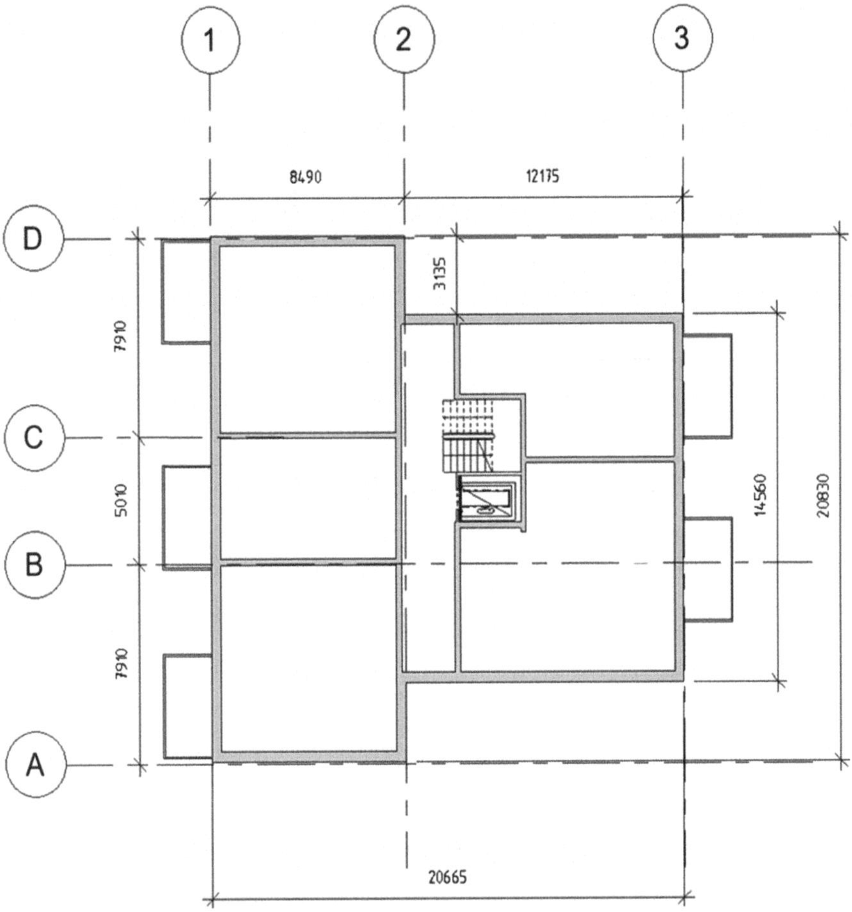

Fig. 5 Load-bearing elements of stories 2–8 (units: mm)

Table 1 Loads applied to building

Load type	Consideration of loads
Self-weight of story slabs	0.5 kN/m^2
Self-weight of roof	0.5 kN/m^2
Imposed load on story slabs	2.5 kN/m^2
Zone load for snow	2.5 kN/m^2
Reference wind speed	23 m/s (terrain type II)

Fig. 6 Modeled building 1

Table 2 Specifications of RC elements utilized in building 2

Type of element	Specification
RC columns	b × h, 300 × 300 mm
RC beams	b × h, 200 × 300 mm
	b × h, 300 × 400 mm
RC slabs	t, 250 mm

When the modeling, analysis, and design of building 2 were completed, the results demonstrated that the building functioned well from the deflection, utilization ratio, and stability perspectives.

3 Results and Discussion

The obtained results from the analysis and design of both buildings are presented and discussed here.

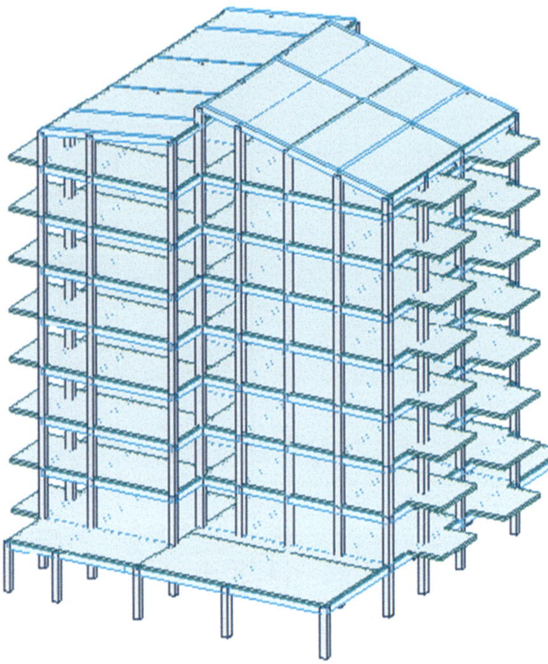

Fig. 7 Modeled building 2

3.1 Stress and Utilization Ratio

High levels of stress in buildings can significantly impact their structure and stability, potentially leading to weakening and even catastrophic failure. Based on the observation, there was a significant difference in stresses between the two buildings due to the type of their structural elements. Building 2 had a more complex structure designed to handle large loads at few points. By concentrating larger loads at specific points, higher stresses can occur, resulting in stress variations across different parts of the building. Building 1 had a more uniform structure and geometry, resulting in an equitable distribution of load throughout the structure. As a result of this uniformity, there were no specific areas designed to carry higher loads. Instead, the entire building was constructed to handle loads equally across its entirety.

The utilization ratio of a structure is a measure of how efficiently its load-carrying capacity is being used compared to its maximum capacity. It evaluates the efficiency of a structure's load-bearing element relative to the applied loads, and if the utilization ratio is below 100%, it indicates that the element can safely carry the loads. In building structures, both structural design and load distribution are crucial for maximizing utilization. A well-planned structural layout with strategically positioned columns and beams enhances efficiency and boosts utilization ratios. Figures 8, 9 and 10 display maximum utilization ratios in buildings 1 and 2. As it can be seen from the figures, the utilization ratios have been achieved below 100%, which

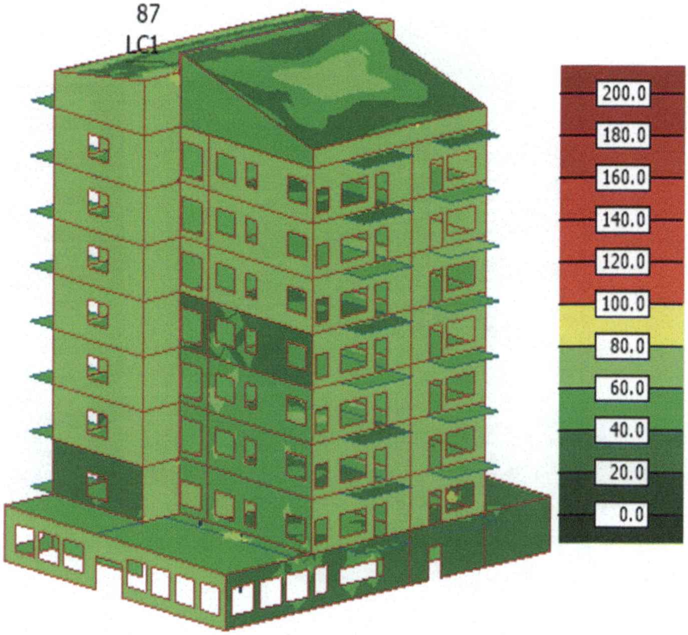

Fig. 8 Maximum utilization ratio (%) for RC slabs and walls in building 1

demonstrates that both buildings functioned well. The figures also depict the load combination (LC) through which these maximum utilizations have occurred.

Moreover, Table 3 outlines the maximum utilization ratios for load-bearing elements of each story of the buildings. The table also shows building 1 reaching a maximum utilization ratio of 87%, whereas building 2 achieves a higher utilization ratio of 99%. Difference in the utilization ratios between the two buildings can depend on several factors, including the structural design of the buildings. It is possible that the RC columns and beams were designed with a more efficient structural layout. By placing them in appropriate positions, they could efficiently carry loads and transfer them to the ground. This optimized structural layout ensured better utilization of structural elements, thereby enhancing overall efficiency.

The distribution of loads can impact the utilization ratio of structural elements. RC slabs and walls are primarily subjected to vertical loads and are utilized to equally distribute the load over a larger area. The design of RC slabs and walls tends to be more conservative to ensure adequate load capacity and stability, often resulting in a lower utilization ratio compared to RC columns and beams. The higher utilization ratio of building 2 was due to RC columns and beams, which offered greater flexibility in interior space design and enabled more optimal area usage. Additionally, these structural elements had higher load-bearing capacities, allowing them to carry more loads and facilitate the construction of taller buildings.

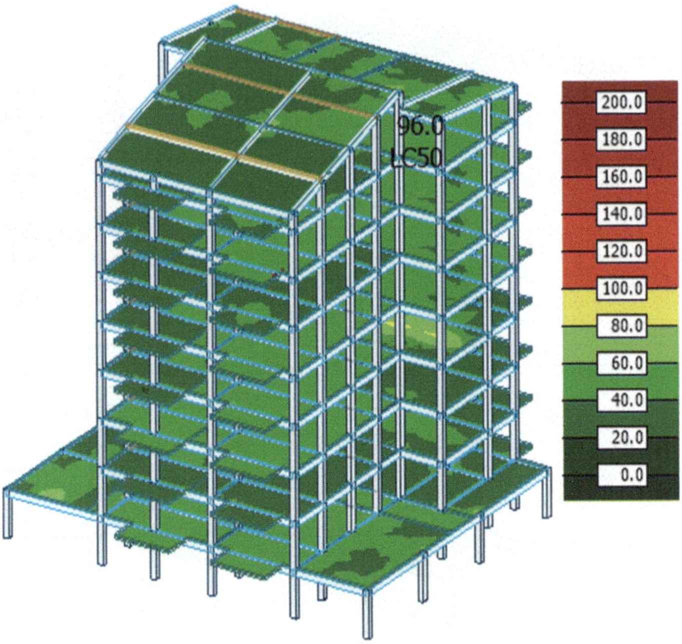

Fig. 9 Maximum utilization ratio (%) for RC slabs in building 2

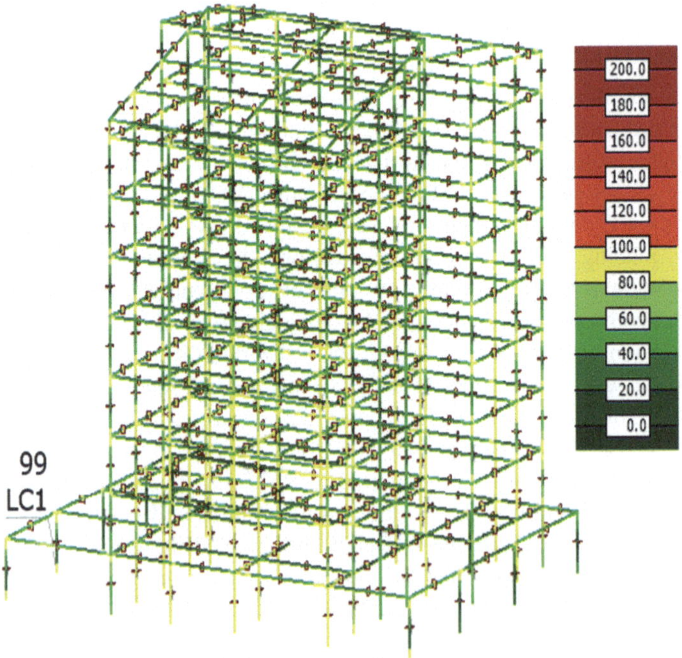

Fig. 10 Maximum utilization ratio (%) for RC columns and beams in building 2

Table 3 Maximum utilization ratios for each story of buildings and for entire buildings

	Building 1		Building 2	
Story	Slab or wall (%)	Column (%)	Slab or wall (%)	Column or beam (%)
Story 1	84	79	96	99
Story 2	87	–	96	99
Story 3	82	–	93	99
Story 4	82	–	92	99
Story 5	87	–	92	99
Story 6	87	–	87	99
Story 7	87	–	96	99
Story 8	87	–	96	99
Roof	87	–	96	99
Maximum utilization ratio for entire building	**87%**		**99%**	

3.2 *Deflection*

The deflections of both buildings were compared. Figures 11 and 12 illustrate the obtained acceptable maximum deflections along with their LCs in buildings 1 and 2, respectively. The maximum deflection in building 1 was 9 mm, with the roof displaying the largest deflection. However, building 2 had the maximum deflection of 7 mm, which was lower than that of building 1. Nonetheless, during the early design stages, it was discovered that the deflections of RC slabs in building 2 exceeded the allowable limit in the serviceability limit state, due to large spans between the columns. To address this issue, additional RC columns and beams were utilized to reduce the spans, ultimately reducing the deflections which led to acceptable values.

3.3 *Reaction Force*

The results of the maximum reaction forces were evaluated for both buildings. The maximum reaction force for building 1 was 563.6 kN, compared to 476.4 kN for building 2. It can be concluded that building 1 had higher reaction forces compared to building 2 due to differences in how the loads were distributed in the structure and the total weight of the buildings.

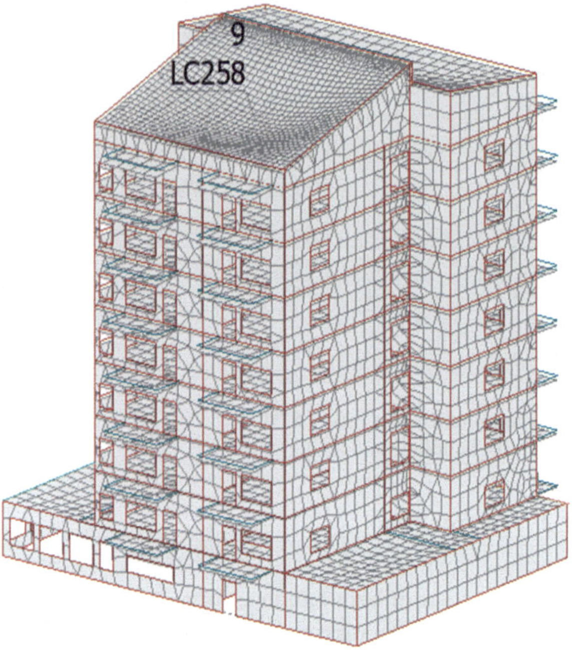

Fig. 11 Maximum deflection in building 1 (unit: mm)

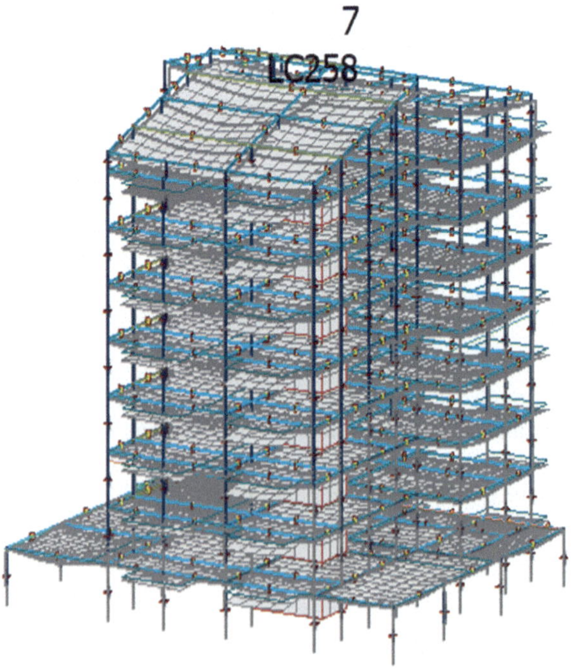

Fig. 12 Maximum deflection in building 2 (unit: mm)

3.4 Stability

A stability analysis was performed on both buildings to identify any deficiencies or risks that could potentially affect their stability and increase the likelihood of structural failure due to the horizontal loads. The analysis was carried out in the second-order analysis that considers the deformed shape of the structures. The results of the analysis indicated that both structures were stable, considering the horizontal loads, and revealed sufficient resistance to such forces. Therefore, no additional strengthening against lateral loads was considered necessary, and both buildings were rigid enough to fulfil the stability requirements. This demonstrated that both buildings were designed in a safe manner and in accordance with the necessary safety standards. In conclusion, the analysis confirmed that both buildings were stable and resilient and could resist both vertical and horizontal forces.

3.5 Weight

The total weight of building 1 was 4169 tons, whereas that of building 2 was 3143 tons, as detailed in Table 4. The analysis conducted on the weight of different building structures revealed that employing RC columns and beams instead of RC walls could reduce the total weight of the building by approximately 25%. This reduction in weight offers several advantages, including lighter foundation loads and a decrease in the requirement for expensive and heavy foundation structures.

The use of lighter materials and building elements also enables more efficient utilization of resources during the construction process and thus lowers the cost. Despite being lighter, these elements still maintain their strength, durability, and ability to meet the functional and structural requirements of the building and can also expedite the construction process.

Table 4 Total weight for each story of buildings and entire buildings

	Building 1	Building 2
Story	Concrete + reinforcement (ton)	Concrete + reinforcement (ton)
Story 1	600	508
Story 2	467	337
Story 3	464	337
Story 4	464	336
Story 5	464	335
Story 6	464	337
Story 7	464	336
Story 8	464	337
Roof	318	280
Total weight (ton)	**4169**	**3143**

4 Conclusions

This study compared two different structural design solutions and their structural performance for a multi-story residential building. A real building as reference building (building 1) located in Sandviken, Sweden, was investigated. Modeling, analysis, and design of building 1 were done using the StruSoft FEM-Design software. Then, load-bearing RC walls of building 1 were replaced with load-bearing RC columns and beams (building 2). Stresses, deflections, utilization ratios, reaction forces, weights, and stability of both buildings were assessed. It was determined that the two solutions functioned well for the building. The results indicated that building 2 with RC columns and beams had a higher maximum utilization ratio compared to building 1 with RC walls. The stresses varied between the buildings due to their structural differences. Furthermore, the results showed that building 1 had lower maximum deflection. Both buildings were stable; however, building 1 proved to be very rigid, as the load-bearing walls acted as shear walls that were very stiff and could withstand lateral loads. Additionally, building 1 experienced larger reaction forces. Based on the results, the use of RC columns and beams in building 2 reduced the total weight of the building by about 25% and decreased the reaction forces, as well. Finally, the use of RC columns and beams allowed for a more open and spacious story plan with larger rooms and fewer obstacles. In addition, it could increase natural light penetration and create a more vibrant environment.

References

1. A. Bahrami, S. Deniz, H. Moalin, Vertical extension of a multi-Storey reinforced concrete building. Int. J. Appl. Mech. Eng. **27**(1), 1–20 (2022) https://www.divaportal.org/smash/get/diva2:1645866/FULLTEXT02.pdf
2. A. Bahrami, F. Ågren, K. Kollberg, Strengthening solutions for deep reinforced concrete beam with cutout opening. Buildings **11**, 370 (2021)
3. Home Improvement Arena, Replacing load-bearing walls, beams, and columns. Hämtad från (2021) http://www.homeimprovementarena.com/2021/09/20/replacing-load-bearing-walls-beams-columns/
4. S.W. Choi, O. Byung Kwan, H.S. Park, Design technology based on resizing method for reduction of costs and carbon dioxide emissions of high-rise buildings. Energ. Buildings **138**, 612–620 (2017). https://doi.org/10.1016/j.enbuild.2016.12.095
5. A. Bahrami, O. Nexén, J. Jonsson, Comparing performance of cross-laminated timber and reinforced concrete walls. Int. J. Appl. Mech. Eng. **26**(3), 28–43 (2021). https://doi.org/10.2478/ijame-2021-0033
6. J. Lindner, Design of steel beams and beam columns. Eng. Struct. **19**(5), 378–384 (1997). https://doi.org/10.1016/S0141-0296(96)00095-8
7. A. Bahrami, D. Azizian, Assessment of glulam and reinforced concrete beams in multi-Storey building. Civil Environ. Eng. **18**(1), 66–75 (2022). https://doi.org/10.2478/cee-2022-0007
8. A. Bahrami, M. Edås, K. Magnenat, J. Norén, The behavior of cross-laminated timber and reinforced concrete floors in a multi-story building. Int. J. Adv. Appl. Sci. **9**(6), 43–50 (2022). https://doi.org/10.21833/ijaas.2022.06.006

9. A. Bahrami, A. Vall, A. Khalaf, Comparison of cross-laminated timber and reinforced concrete floors with regard to load-bearing properties. Civil Eng. Arch. **9**(5), 1395–1408 (2021). https://doi.org/10.13189/cea.2021.090513
10. A. Zona, G. Ranzi, Finite element models for nonlinear analysis of steel– Concrete composite beams with partial interaction in combined bending and shear. Finite Elem. Anal. Des. **47**(2), 98–118 (2011). https://doi.org/10.1016/j.finel.2010.09.006
11. S. Navaratnam, T. Ngo, T. Gunawardena, D. Henderson, Performance review of prefabricated building systems and future research in Australia. Buildings **9**(2), 38 (2019). https://doi.org/10.3390/buildings9020038
12. A. Baghchesaraei, M.V. Kaptan, O.R. Baghchesaraei, Using Prefabrication Systems in Building Construction. **10**, 24 (2015) https://www.researchgate.net/publication/296639797
13. A. Kaveh, R.A. Izadifard, L. Mottaghi, Optimal design of planar RC frames considering CO2 emissions using ECBO, EVPS and PSO metaheuristic algorithms. J. Build. Eng. **28**, 101014 (2020). https://doi.org/10.1016/j.jobe.2019.101014
14. I. Laníková, P. Štěpánek, P. Simunek, Optimized design of concrete structures considering environmental aspects. Adv. Struct. Eng. **17**(4), 495–511 (2014). https://doi.org/10.1260/1369-4332.17.4.495
15. A.R. Kulkarni, V. Bhusare, Structural optimization of reinforced concrete structures. Int. J. Eng. Res. V5 **07**, IJERTV5IS070156 (2016). https://doi.org/10.17577/IJERTV5IS070156
16. M.J. Esfandiary, S. Sheikholarefin, H.A. Rahimi Bondarabadi, A combination of particle swarm optimization and multi—Criterion decision—Making for optimum design of reinforced concrete frames. Int. J. Optim. Civ. Eng. **6**(2), 245–268 (2016)
17. M. Leps, Multi-objective optimization of reinforced concrete frames, Rio de Janeiro, Brazil, 30 May–03 June 2005
18. G.H. Kusuma, J. Budidarmawan, A. Susilowati, Impact of concrete quality on sustainability. Procedia Eng. **125**, 754–759 (2015). https://doi.org/10.1016/j.proeng.2015.11.122
19. G.F. de Medeiros, M. Kripka, Optimization of reinforced concrete columns according to different environmental impact assessment parameters. Eng. Struct. **59**, 185–194 (2014). https://doi.org/10.1016/j.engstruct.2013.10.045
20. M. Ženíšek, J. Pešta, M. Tipka, V. Kočí, P. Hájek, Optimization of RC structures in terms of cost and environmental impact—Case study. Sustain. For. **12**(20), 8532 (2020). https://doi.org/10.3390/su12208532

Innovative Materials and Strategies for Low-Carbon Concrete

Alireza Bahrami, Muhammad Hammad, and Shafqat Ullah

1 Introduction

The production of cement and concrete is considered a major source of greenhouse gas emissions, demanding an urgent need for decarbonization initiatives [1, 2]. The processes of grinding and transportation of the material, as well as carbon dioxide (CO_2) released during cement production, are the primary causes of the environmental challenges [3]. The use of concrete, which mostly utilizes cement, intensifies these environmental issues. Decarbonizing the manufacturing of ordinary Portland cement (OPC) and concrete is therefore necessary for climate change mitigation and sustainable building practices [4]. The sustainable climate objectives in terms of challenges can be achieved by the reduction of CO_2 release during manufacturing of OPC and concrete. Furthermore, decarbonization can significantly improve sustainability in the construction industries as well as establish long-term environmental preservation. The development of cementitious materials is the primary cause of the large carbon footprint associated with conventional concrete production. But as more eco-friendly options become available, the paradigm is changing and placing a higher priority on the environmental sustainability without losing its functionality [5, 6]. By the substitution of several supplementary cementitious materials (SCMs) such as fly ash (FA), slag, or silica fume (SF), these

A. Bahrami (✉)
Department of Building Engineering, Energy Systems and Sustainability Science, Faculty of Engineering and Sustainable Development, University of Gävle, Gävle, Sweden
e-mail: alireza.bahrami@hig.se

M. Hammad
School of Civil Engineering, University of Nottingham, University Park, Nottingham, UK

S. Ullah
Department of Civil Engineering, University of North Dakota, Grand Forks, ND, USA

A. Bahrami (ed.), *Advances in Sustainable Concrete for Construction*,
https://doi.org/10.1007/978-3-031-85052-3_3

sustainable concretes not only utilize less cement but also make use of industrial by-products, which increases resource efficiency. A study has demonstrated that using high-volume FA in concrete, that is, half substitution of OPC, reduces shrinkage by 31% in comparison to conventional concrete [7]. Further enhancing concrete's eco-credentials is the incorporation of recycled aggregate (RA) from demolished structures, which reduces the dependency on natural resources and landfill space [8].

The idea of carbon sequestration in concrete matrices is one of the most innovative advancements in sustainable construction [9, 10]. Ordinary concrete absorbs CO_2 over time through a process called carbonation. A number of solutions have been proposed for decreasing carbon emissions and stopping climate change, such as directly injecting CO_2 into concrete mixtures or using CO_2-absorbing binders. Research presented that cement specimens that were cured in the air for 20 h had lower hydration and carbonation products compared to specimens exposed to CO_2. Moreover, 8% of carbon in cement was captured by the carbonation process, resulting in a 16% degree of carbonation [11]. Besides the addition of eco-friendly binders and fillers in concretes, several other approaches have been explored recently to reduce, capture, and utilize carbon [12]. Additionally, mixed design optimization frameworks and machine learning models have been developed recently to select suitable and optimum mixed designs for sustainable concrete without wasting time, cost, and energy [13, 14].

This chapter critically reviews the current developments in eco-friendly concrete, carbon sequestration, and usage for sustainable construction against the backdrop of rapid innovation and transformational potential. Furthermore, this study attempts to clarify the current critical obstacles and future directions in the search for sustainable construction materials and practices by synthesizing existing literature, analyzing significant technological breakthroughs, and identifying research gaps. The layout of this chapter is illustrated in Fig. 1.

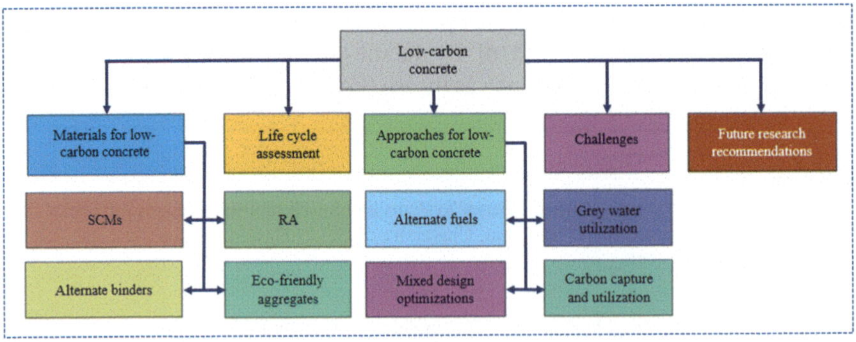

Fig. 1 Flowchart of topics discussed in this chapter

2 Materials for Low-Carbon Concrete

Low-carbon concrete (LCC) is becoming more popular as a sustainable concrete compared to conventional concrete, because of its less environmental impacts as well as less carbon footprints. One possible approach to achieve LCC is to use cementitious materials that replace or supplement the primary binder in conventional concrete. SCMs can add noticeable strength, durability, and other mechanical properties to concrete structures. The different construction materials that can be used for LCC are explained below.

2.1 SCMs

SCMs are an inorganic material that can be partially utilized in concrete replacing OPC, which enhances the hydration reaction as well as the hydration product [15]. For a decade, a large number of SCMs have been employed in concrete to achieve different objectives such as recycling waste products, energy consumption degradations, cost reduction per unit volume as well as enhancing the quality of concrete in terms of compressive strength (f'_c), and durability as well as improving mechanical characteristics. These SCMs are valuable in enhancing the performance of concrete through hydration or pozzolanic reaction, improving the workability of fresh concrete, f'_c, and durability of concrete [16]. The concepts of industrial ecology are best displayed by blended cements, which contain SCMs that are usually made from industrial by-products. As the requirement for cement continuously increases, the utilization of alternative SCMs obtained from local materials becomes important [17]. Nevertheless, this introduces challenges, as alternative SCMs often exhibit smaller volumes and more variable compositions. Consequently, there is a pressing need for more effective and appropriate methodologies to determine the performance and impact of SCMs in concrete applications. Table 1 summarizes the different SCMs used in concrete.

2.1.1 FA

FA can be obtained from the residue waste after the coal-burning process in thermal power plants, and it is reported that over 1 billion tons of waste products are produced yearly [19]. FA can be efficiently incorporated in concrete in place of OPC. The amorphous silica in FA after reacting with calcium hydroxide ($Ca(OH)_2$) forms calcium–silicate hydrate [20]. Additionally, FA increases the workability and ultimate f'_c of concrete [21]. Research has indicated that using high-volume FA in concrete, that is, replacing 50% of OPC, reduces shrinkage by 31% in comparison to conventional concrete [22]. FA particles typically have a spherical shape and can be solid or hollow from the inside, primarily consisting of glassy (amorphous)

Table 1 Different SCMs, sources, and their manufacturing process [18]

Source	SCMs	Abbreviation	Process
Industrial by-product	Fly ash	FA	A by-product of coal combustion
	Ground granulated blast furnace slag	GGBS	Waste from pig iron production
	Silica fume	SF	A by-product of metallic and iron–silicon alloys
	Copper slag	CS	Waste from copper processing
	Ferrochrome ash	FeA	Dust from ferrochrome production unit
	Lime sludge	LS	By-products from different industries such as paper industries and water softening
	Red mud	RM	Waste produced from alumina refinery plants
Natural SCMs	Wollastonite	WS	Formed naturally by interacting with silica and limestone in volcanoes
	Marble powder	MP	Slurry waste obtained from stone formation
	Calcined clay	CC	Calcination of clay that includes minerals such as metakaolin and montmorillonite
	Metakaolin	MK	Calcining kaolin clay at temperature from 750 to 800 °C
Bio-ash	Rice husk ash	RHA	By burning rice husk
Wastes	Glass powder, eggshell, and ceramic waste powder	GP, ESP, and CWP	Discarded glass, dried eggshells, and waste produced by ceramic industries

material, with particle sizes ranging from less than 1 to 150 μm [23, 24]. Scanning electron microscope (SEM) micrographs of FA are depicted in Fig. 2. In addition to its environmental benefits, such as waste disposal and CO_2 reduction, FA reduces the heat of hydration and cracking due to heat during the initial stages and improves the performance of concrete, particularly in later stages [25]. According to Woolley et al. [26], the pulverized FA are classified as low- and high-lime FA, and the categorization roughly refers to FA classes F and C according to ASTM, respectively. The elemental oxide contents in FA classes F and C can vary considerably, as outlined in Table 2.

Several researchers have examined the partial replacement of FA and their effects on f'_c of concrete [24, 28–31]. Barbuta et al. [31] reported that the early age f'_c of concrete decreased with the incorporation of FA, where higher FA content as cement replacement had a significant reduction in f'_c. Samples without fibers showed f'_c decrement of 11%, 30%, 25%, 34%, and 60% with FA contents of 10%, 15%, 20%, 30%, and 40%, respectively, compared to the control specimens. This trend was similarly observed in specimens with fibers. Moreover, Gencel et al. [32] evaluated the effects of FA on composite characteristics using various percentages of FA

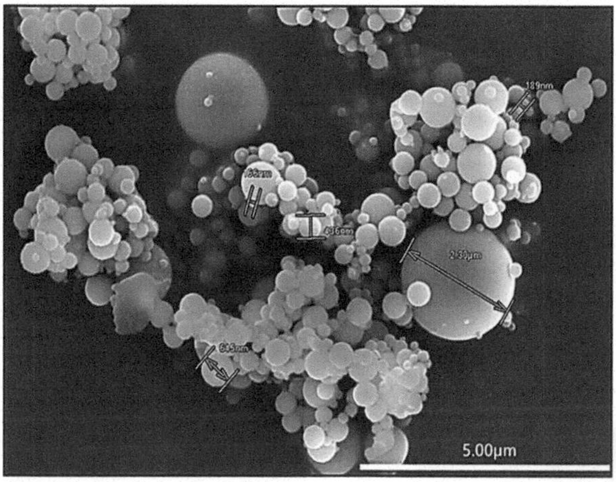

Fig. 2 Micrograph of FA 2 µm [27]

Table 2 Element oxides in FA classes F and C

S. no.	Element oxide	Class F	Class C
1	SiO_2	38–62	11.8–46.4
2	Al_2O_3	16.5–35.5	2.6–20.5
3	Fe_2O_3	2.7–21.3	1.4–15.6
4	CaO	0.5–14.2	15.2–55.1
5	MgO	0.3–5.1	0.1–6.8
6	SO_3	0.02–4.81	1.5–13
7	Na_2O	0.1–3.5	0.2–3
8	K_2O	0.1–4.1	0.3–9
9	TiO_2	0.5–2.5	0.5–1.0
10	P_2O_5	0.1–1.8	0.2–0.45
11	MnO	0.03–0.12	0.25–0.3
12	LOI	0.3–33.1	0.3–12

substitutions for OPC, resulting in decreased f'_c compared to the control mix, particularly evident when FA concentration was high. Huang et al. [33] examined the influence of two types of FA categorized by loss on ignition (LOI), low (4.5%) and high (7.9%) LOI. They found that employing (4.5%-LOI) FA at a lower ratio increased f'_c, with the maximum enhancement noted at 40% FA content, demonstrating 16.9% higher f'_c than the control sample. However, the decrease in f'_c was found to be associated with increased FA proportions due to the smaller grain size of (4.5%-LOI) FA, leading to a denser and more compact microstructure. Conversely, the reduction in f'_c was noted with high (7.5%-LOI) FA addition due to larger grain size and reduced pozzolanic reaction. Furthermore, f'_c of composites with more FA substitution illustrated improvement at a later stage compared to the control sample due to slow pozzolanic reaction.

Moreover, several researchers have studied the effects of FA addition in self-compacting concrete. For instance, replacing 30% OPC with FA led to a 10%

decrement in f'_c, while replacing 50% FA resulted in a 25% decrement compared to the control specimen [34, 35]. Similarly, a decrement of nearly 45% and 36% was observed in mixes with half FA substitution. In addition, a 70% FA substitution in self-compacting concrete indicated a significant decrement of nearly 65% and 45% after curing for 7 and 90 days, respectively [36].

2.1.2 SF

SF is a residue obtained from silicon metal and ferrosilicon alloy and is mostly used partially in place of OPC in concrete [37]. This SCM has a high amount of silicon dioxide in amorphous form and high pozzolanic activity. The particles of SF have a shape like a sphere with a diameter ranging between 0.1 and 0.2 μm, as displayed in Fig. 3.

The specific surface area of SF varies between 13,000 and 30,000 m^2/kg, which contributes to the high reactivity of the pozzolanic reactions that enables them to react quickly with ($Ca(OH)_2$) generated during OPC hydration. The physical and chemical characteristics of SF are listed in Tables 3 and 4, respectively.

Various researchers have studied the impacts of SF on several characteristics of concrete including its f'_c, durability, abrasion resistance, flexure, and tensile strength [39–42]. Half substitution of SF significantly affects f'_c of concrete in early ages, and it acts as a filler and nucleation site that can aid in making the denser microstructures and thus increasing the mechanical properties of concrete [43]. Furthermore, SF enhances concrete durability by lowering the permeability of concrete as well as increasing the resistance to chemical and physical degradation mechanisms [44]. Incorporating SF in concrete remarkably improves the flexural and tensile strengths as well as increases the resistance to cracking under tensile and bending loadings [45]. Another study assessed the impacts of various proportions of SF on f'_c of concrete. The authors concluded that an increment in the content of SF increased the ultimate f'_c of concrete. There was an increment of 18% in f'_c of

Fig. 3 SEM picture of SF with a diameter of 0.1 μm [38]

Table 3 Physical characteristics of SF [38]

S. no.	Properties of SF	Values
1	Particles size	<1 μm
2	Bulk density	132–430 kg/m^3
3	Slurry	1310–1440 kg/m^3
4	Density	490–720 kg/m^3
5	Specific gravity	2.20
6	Surface area	13,000–30,000 m^2/kg

Table 4 Chemical properties of SF [38]

S. no.	Oxides	Weight (%)
1	SiO_2	92.26
2	Al_2O_3	0.89
3	Fe_2O_3	1.97
4	CaO	0.49
5	MgO	0.96
6	K_2O	1.31
7	Na_2O	0.42
8	SO_3	0.33
9	LOI	1.37

concrete with 15% SF contents after 55 days when compared to the control concrete and an increase of 14% when compared to concrete employing 10% SF. They also evaluated the influence of SF on the split tensile strength, chemical durability, and sorptivity. It was noticed that while there were no significant effects of SF on the split tensile strength, increasing SF content led to an increase in the resistance against chemical attacks. Furthermore, it was concluded that the sorptivity decreased after incorporating SF in a mix [42].

2.2 Alternatives for Binders and Clinker

To reduce the dependency on OPC clinker as the main binder in concrete, researchers and industry professionals have been testing various alternative materials in response to the environmental concerns raised by the preparation of OPC. The goal of these substitutes has been to lessen energy usage, as well as the carbon footprint of concrete manufacturing, and advance sustainability in the building industry.

2.2.1 CC

CCs are natural pozzolanic materials that are heated to high temperatures (generally between 600 and 800 °C) [46]. Limestone CC cement (LC3) transfers equivalent mechanical properties to ordinary concrete mixes while significantly lowering CO_2 emissions and energy consumption by using limestone and CC as clinker substitutes [47]. LC3 has become more popular because of its ability to encourage

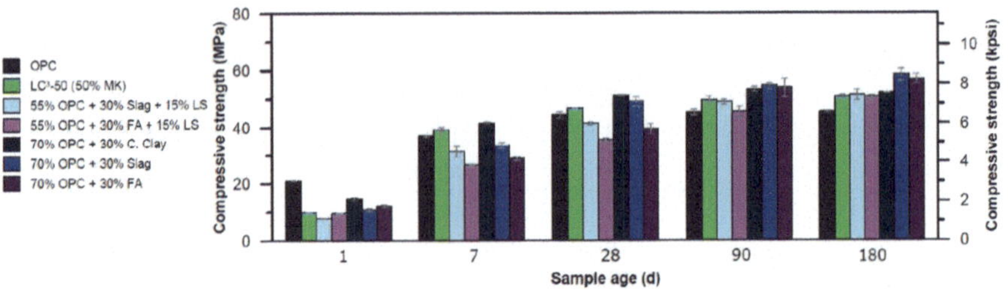

Fig. 4 Development of f'$_c$ in LC3 compared to other cement and SCM blends [46]

sustainable growth and lessen the global environmental effect of construction activities.

CCs lower the total carbon footprint and energy usage associated with the cement manufacturing while enhancing the workability, durability, and early strength development [7, 46, 47]. The strength development in LC3 compared to other SCMs-OPC is depicted in Fig. 4. Joseph et al. [48] examined the mechanical characteristics incorporating CC in concrete. They compared f'$_c$ and tensile strength of concrete incorporating CC, and it was observed that the concrete could attain equivalent or even higher strength levels than an OPC system when incorporating up to 30% CC or a blend of 50% CC and limestone fines in an LC3 system. Sabir et al. [49] documented the usage of MK and CC as pozzolans for concrete, highlighting numerous advantageous impacts of MK inclusion on pore structure and properties of concrete. Up to 25% MK was utilized, as a partial substitution for the cement-enhanced chloride resistance, evidenced by a lower diffusion coefficient. Moreover, an increment in the diffusion reduction factor was observed with higher levels of replacement. Dhandapani et al. [50] experimentally investigated the impacts of LC3 on the mechanical characteristics, shrinkage, and durability of various concrete mixes. Concrete produced using LC3 displayed notably higher resistivity compared to mixtures made with OPC and FA.

2.2.2 Alkali-Activated Materials

Alkali-activated materials (AAMs) are becoming more popular as eco-friendly and sustainable construction materials because they help minimize CO_2 emissions and make use of industrial by-products [51]. Nonetheless, their practical use faces challenges due to the difference in their inherent characteristics during the initial stages, impacting their handling, placement, and early-age performance [52]. AAM represents a broad category comprising SCMs made by the interaction of alkaline activators with different substrates. Among these, geopolymer particularly is a subtype of AAMs [53, 54]. Research has demonstrated that AAMs can achieve mechanical properties comparable to or even surpass those of materials based on OPC. This characteristic renders them more eco-friendly than conventional concrete [29]. Several researchers studied the durability and mechanical characteristics of AAMs and OPC-made concretes under different chemical attacks. Wang et al. [52] examined the effects of AAMs compared to OPC under chemical attacks. They concluded that AAMs showed better performances in terms of durability compared to OPC.

The literature review on durability indicated that AAMs made from MK do not generate sulfate–aluminate minerals like ettringite during the hydration process, leading to improved resistance to sulfate attack in calcium-free or low-calcium AAMs [30, 55, 56]. Tao et al. [57] determined sulfate resistance by calculating the strength retention rate and found that MK-based AAM had substantially higher strength preservation rates than OPC in sodium sulfate solution. Additionally, Hou et al. [28] noted differences in f'_c and density of FA-employed AAM after 155 days of submersion in different sulfate solutions, suggesting that the impact of sulfate solutions on the properties of FA-employed AAM was associated with diffusion.

2.3 *RA and Concrete*

The last decade has witnessed an increase in value and support for sustainable approaches in concrete by reducing the usage of natural materials like cement and aggregates to lower the rate of natural resource depletion. This practice contributes to a cleaner environment and fosters sustainability by utilizing recycled materials in concrete production [8, 12]. The recycling and partial replacement of each ingredient in concrete with several by-products has been thoroughly evaluated in numerous studies [58–60]. Additionally, Llanes et al. [61] highlighted that reusing construction and demolition concrete waste could yield substantial net benefits compared to the procurement and manufacturing of new materials for aggregates. Currently, aggregate products from most recycling units include granular fills, subbase, and recycled concrete aggregate (RCA), which is typically mixed with other ingredients. Companies also utilize stone, concrete, brick, asphalt, and clay, in low-strength concrete for low-cost trench fill material [62, 63]. RA concrete consists of crushed and coarser waste concrete, comprising nearly 90% of the concrete weight, with generally total impurity levels lower than 2% of the total mass [64]. Meanwhile, as landfills become increasingly scarce, this practice not only lessens the demand for landfill space but also alleviates the project's financial burden. By addressing both waste management and novel material manufacturing requirements, the transportation needs for the tasks are markedly reduced. Beyond resource management, RCAs play a role in CO_2 absorption [36]. Wang et al. [65] studied the impacts of various aggregate treatments on RA and recycled concrete qualities. Various parameters including f'_c and elastic modulus (E) of five different kinds of recycled concrete were assessed. The findings exhibited that while the slurry wrapping method has increased the resistance to chloride ion penetration, carbonation significantly improved mechanical characteristics and dry shrinkage resistance. While the slurry wrapping affected both old and new interfacial transition zones differentially, carbonation had a greater effect on older interfacial transition zones than new ones. Different results of recycled concrete after testing are illustrated in Fig. 5. As shown

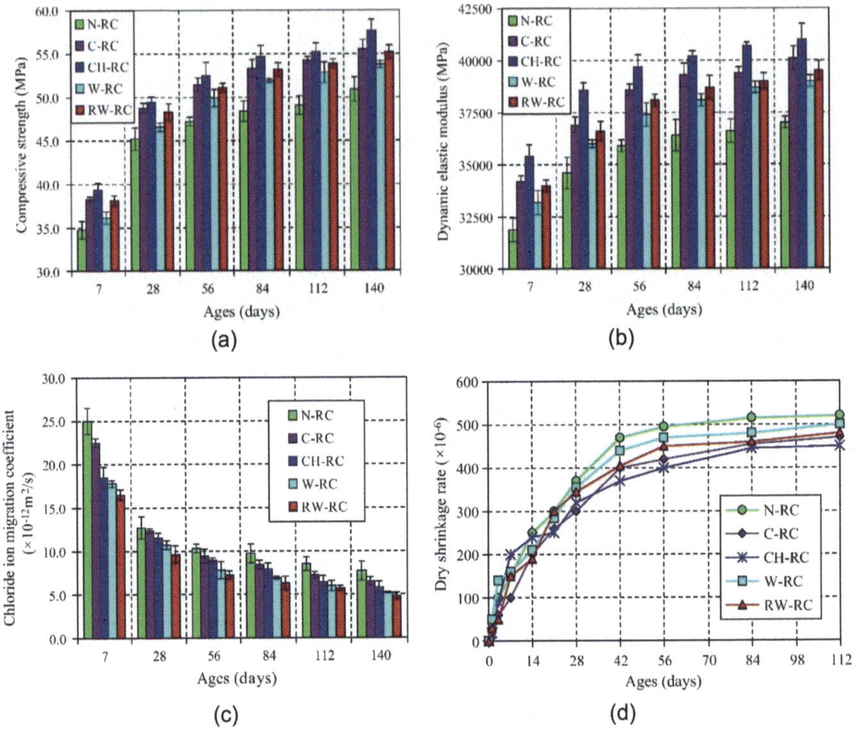

Fig. 5 Results of tested recycled concrete: (**a**) f'$_c$, (**b**) dynamic E, (**c**) chemical durability, and (**d**) dry shrinkage [65]

in Fig. 5, N-RC, C-RC, CH-RC, W-RC, and RW-RC represent the control specimen, carbonation method with high CO_2 concentration, carbonation method with $Ca(OH)_2$ emulsion, ordinary wrapping method, and new paste wrapping method, respectively.

Alqarni et al. [66] investigated the effects of maximum aggregate sizes (10 mm and 20 mm), different RA replacements (0%, 33%, 67%, and 100%), and three treatment approaches on the characteristics of concrete. The outcomes revealed that treated RA improved concrete slump by 15–35%, while untreated RA had a detrimental effect on fresh concrete characteristics, particularly at higher replacement levels. Furthermore, treated RA decreased the sorptivity, which was dominant with aggregate sizes over 25 mm, and increased f'$_c$ independent of aggregate size. Besides experimental work, many machine learning frameworks have been trained by various researchers to anticipate the performance of waste and RCA-based concrete [67–70].

2.4 Eco-Friendly Aggregates

Both finer and coarser aggregates are the main filler ingredient in concrete except for engineered cementitious composites, which do not contain coarse aggregates [71]. A growing focus in peruse for sustainable construction methods is the use of eco-efficient aggregates made from reused materials, industrial by-products, or renewable resources [58, 60, 64]. Multiple environmental benefits are provided by these eco-friendly aggregates, such as lesser carbon footprint, retention of natural resource utilization, and decrement in waste production [6]. Expanded clay, shale, or slate are examples of lightweight aggregates that are manufactured from natural materials or industrial leftover. These aggregates are perfect for reducing the weight of concrete because of their low density and strong strength-to-weight ratio. In addition to improving the thermal insulation and acoustic properties, lightweight aggregates can save energy during construction and transportation [72]. Hussain et al. [73] evaluated the possibility of manufacturing artificial lightweight expanded clay aggregates (ALECA) and investigated the impact of admixture amounts, thermal value, and size of pellets on the mechanical properties of ALECA. After studying five clay fields, ALECA was created using different mix formulations, as illustrated in Fig. 6. Nandipur clay had optimum bloating properties, which made it suitable for ALECA production. In addition to saving a lot of money and reducing carbon footprints, these lightweight aggregates help preserve natural resources in the building industry. Furthermore, they also assessed the application of coal combustion residue (CCR). Notable improvements were seen when CCR and kerosene were combined as a bloating agent. Density ranged between 0.4 and 0.95 g/cm^3, while particle strength increased from 0.4 to 8.5 MPa. The final aggregates had perfect strength-to-density ratios that were appropriate for a range of concrete uses, suggesting that CCR in ceramic aggregates are suitable to support waste management and sustainable building methodologies [74].

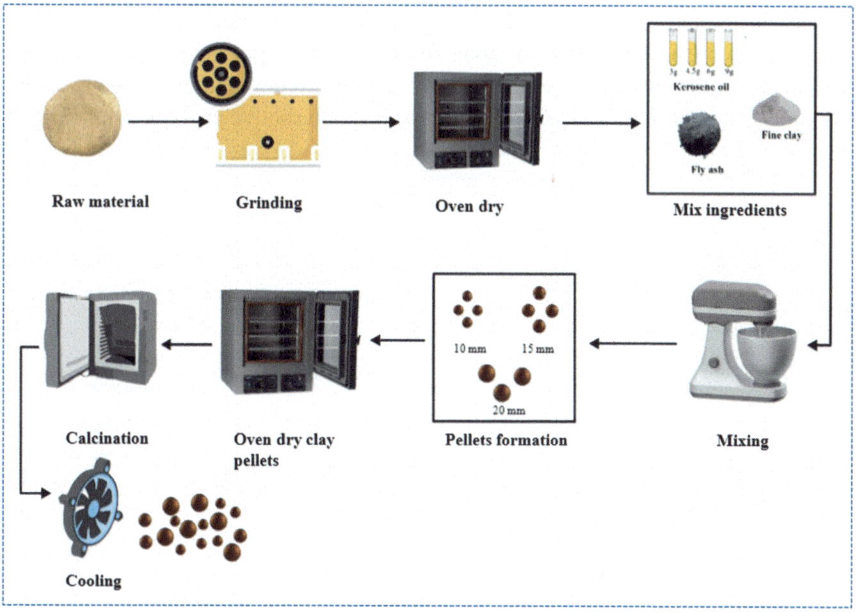

Fig. 6 Development process of artificial sustainable lightweight aggregates [74]

3 Approaches for Reduction of Carbon Footprints

Besides the replacement of cement with SCMs, various approaches are being used worldwide to tackle the issue of CO_2 emissions. These approaches include the use of treated wastewater, use of alternative fuels to provide energy to concrete production processes, optimization of the concrete mix designs through machine learning and a few other frameworks, carbon capture and utilization, etc. Some of these approaches are presented in the following sections.

3.1 *Alternative Fuels*

The cement production process mostly employs fossil fuels to provide energy [75]. Therefore, this process contributes to approximately 5% of CO_2 emissions worldwide [76]. Reports indicated that manufacturing of 1 kg of clinker results in the release of nearly 1 kg of CO_2 into the atmosphere [75]. Specifically, the calcination process within cement kilns is responsible for emitting around 0.55 kg of CO_2 per kg of cement clinker [77]. This requirement for higher energy production during the cement manufacturing process is a major concern for environmental protection agencies. The typical fossil fuels mainly come from coal, pet coke, and natural gas. All these fuel sources are detrimental to the environment. Several research works have been conducted on using potentially eco-friendly and sustainable alternatives [78].

Waste materials, like automobile shredder residue, carpet remnants, textiles, wax residue, meat and bone leftovers, landfill gases, and fuels obtained from municipal solid wastes, are utilized as alternative fuels in cement production [79, 80]. The ingredient percentage of different alternative fuels used for manufacturing of cement is demonstrated in Fig. 7 [81]. By incorporating alternative fuels with properties

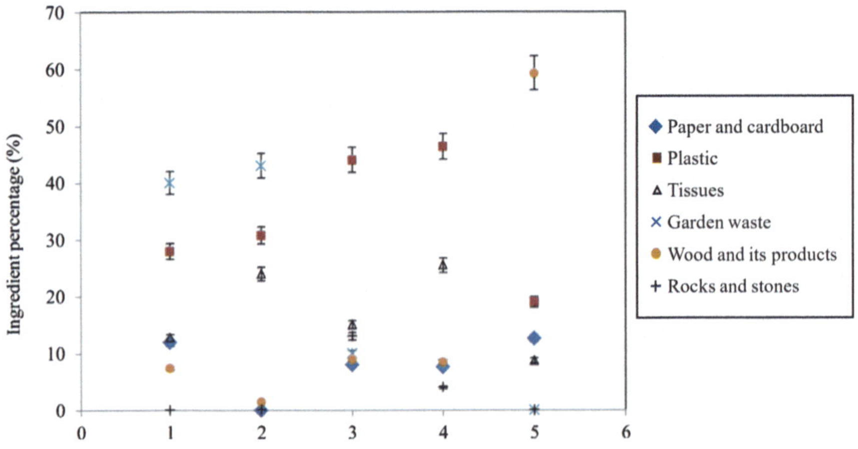

Fig. 7 Ingredient percentages of different alternative fuels for cement production [81]

akin to pet coke, cement manufacturers can maintain operational efficiency and meet environmental regulations while reducing reliance on fossil fuels. Further studies have shown that waste oil, fabrics, waste plastic, and sludge can be employed as a substitute fuel source for OPC manufacturing. The emission of CO_2 in using these fuel types is very low compared to fossil fuels, but pollution caused by their burning remains a threat to the environment.

Another potential energy source is biomass, which is obtained from burning agricultural waste such as wood and grass. Biomass energy is considered as renewable energy due to its generation from agricultural wastes, which are continuously produced. Moreover, it can be considered as a carbon-neutral source of fuel, with the assumption that the CO_2 released during its utilization is approximately equal to the amount a tree takes up to grow [82]. Besides its benefits in sustainability compared to other alternative fuels in the concrete industry, biomass energy has an extra advantage. When biomass is combusted, it produces waste that contains high levels of silicate and aluminate. These waste materials can be reused as SCM, effectively replacing OPC. This not only decreases the adverse impact on the environment but also enhances the efficiency of resource utilization in concrete production.

3.2 Gray Water Utilization

The widespread use of concrete in various infrastructure projects consumes around 1 billion cubic meters of water, ranking it as the foremost consumer of freshwater worldwide [83]. This substantial water usage is accompanied by the generation of considerable wastewater within the industry. Particularly in landlocked regions and countries with restricted access to freshwater, concrete production poses a notable threat to water reserves. Moreover, the mobilization of water from its origin to the project location in these areas further amplifies environmental concerns by contributing significantly to CO_2 emissions.

One replacement to conventional mixing methods includes utilization of gray water, which is obtained locally from cleaning concrete mixers [84]. The process of cleaning these mixers generally takes around 1000 liters of water, leading to millions of cubic meters of gray water being generated per annum globally. This gray water can be employed for mixing of concrete ingredients, reducing CO_2 emissions. Reclaiming and reusing gray water is considered as having a large impact, not only on addressing current water shortage challenges but also on enhancing a country's environment and economy [85]. The influence of recycled household wastewater on f'_c of concrete was examined by Shekarchi et al. [86]. Multiple combinations of mixing and curing were studied with three kinds of wastewater. The findings exhibited that whereas secondary treated wastewater decreased f'_c up to 10% when compared to the control specimen, tertiary treated wastewater increased f'_c by 7%, as depicted in Fig. 8. Wastewater with primary treatment had a negligible impact on strength. When samples were blended or cured with primary and secondary wastewaters for up to 6 months, f'_c is noted to increase by up to 17%. On the other hand,

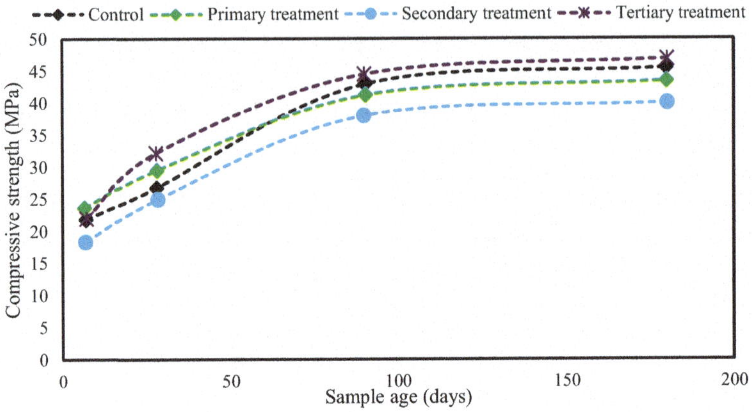

(a)

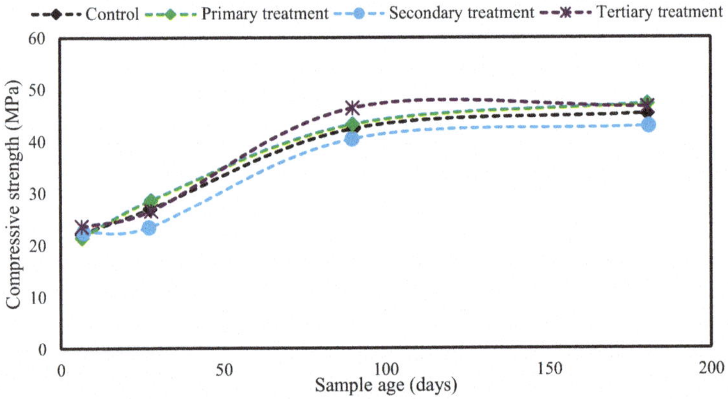

(b)

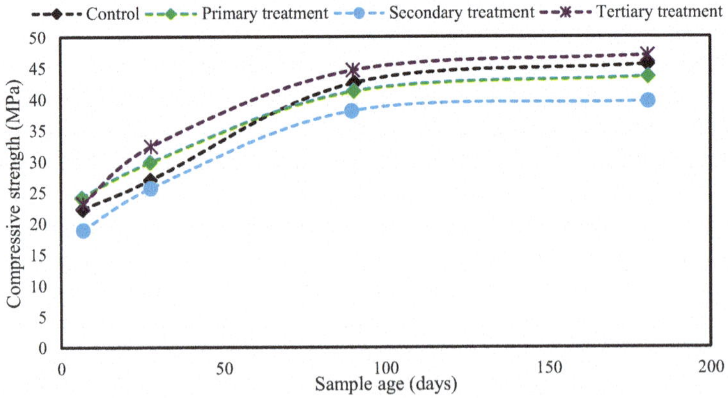

(c)

Fig. 8 (**a**) Tap water for curing and three kinds of treated waters for mixing, (**b**) tap water for mixing and various kinds of treated waters for curing, and (**c**) various kinds of treated waters for both mixing and curing [86] the strength declined by 9–18% throughout the same period while using secondary treated wastewater, as illustrated below.

On comparing concrete mixed with drinkable water, Mane et al. [87] reported that the usage of treated sewage water in concrete enhanced its f'_c by 10.83%. This wastewater treatment can also be employed for curing because the strength change is minimal.

Furthermore, Jabri et al. [87] stated that wastewater had no noticeable impact on high-strength concrete (HSC). They compared the outcomes for mixtures that replaced 25%–100% of wastewater. At 7 and 28 days for the 50% wastewater replacement, the strength was similar to the control mix. The abovementioned studies confirm that well-treated wastewater can be used as a replacement for fresh water.

3.3 *Optimized Mixed Designs*

Prior studies have demonstrated that when adding waste-based aggregates to concrete mixes, extra binders are often necessary to preserve mechanical characteristics. Incorporating additional binders may raise global warming potential (GWP), which poses a difficulty in creating "green" or eco-friendly concretes. Nevertheless, using waste-based aggregates offers a solution to prevent further utilization of natural resources [88]. Therefore, various researchers have developed mixed design optimization frameworks and tools to select optimized mixed designs for concretes, containing different types of waste materials and by-products [67, 68, 70]. For instance, Malik et al. [89] trained an artificial neural network (ANN)-based machine learning framework to predict the complete stress–strain curves of the engineered cementitious composites. The optimization of the composite reduced the time, cost, and effort required to select a suitable mix design. Shobeiri et al. [14] made a framework to optimize concrete mix design that employs several waste-based aggregates, such as rubber, plastic, glass, and slag. It uses the genetic algorithms to reduce the use of natural resources and GWP. With accurate predictions from the ANN model, the framework remained effective in decreasing CO_2 emissions by 50% when compared to equal-strength mixes in the database. Campos et al. [13] presented a new approach for mix design of sustainable HSC that utilizes less cement and emits less CO_2 than current methodologies. The suitable ratios of sand and aggregates were chosen using particle packing method. For validation, HSC was manufactured, utilizing a mixture of fine and coarse aggregates, silica fume, and rock powder. The results displayed that f'_c of 104 MPa and E of 60 GPa were obtained with just 336 kg/m^3 of cement. In comparison to the values mentioned in the literature for HSC, the approach resulted in the cement utilization of only 3.2 kg per 1 MPa of f'_c, as shown in Fig. 9. In Fig. 9, Vp is the concrete paste volume, whereas 29, 31, 33, and 35 are the volume percentages.

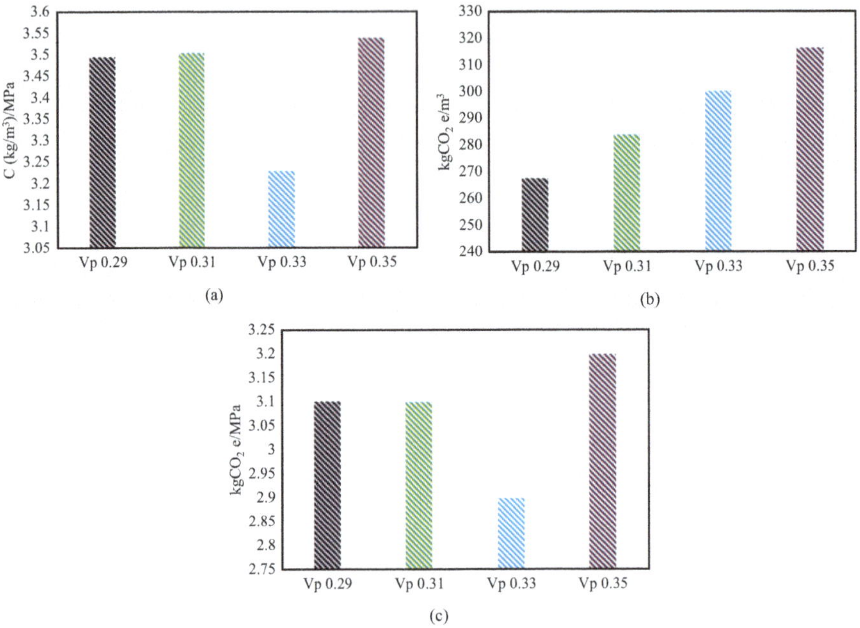

Fig. 9 (**a**) OPC required for 1 MPa of f'$_c$, (**b**) CO_2 emission per m^3 of concrete, and (**c**) CO_2 emission to obtain 1 MPa of f'$_c$ [13]

3.4 Carbon Capture and Utilization

Manufacturing of cement needs a large amount of energy and resources, and it is the reason for nearly 10% of total emissions of greenhouse gases as CO_2 [4, 80]. To lower CO_2 emissions, the cement industry is constantly researching and developing low-carbon and energy-saving solutions [90]. Furthermore, by using recently developed carbon capture and utilization technology, CO_2 emissions during the production of OPC can be reduced effectively [12, 91, 92]. Instead of simply sequestering it, carbon capture and utilization processes convert captured CO_2 into valuable products, treating it as a renewable carbon resource. Unlike carbon capture and storage (CCS), which raises questions about long-term sequestration impacts, carbon capture and utilization provides a more effective solution. Apart from the substantial benefits over CCS, the conversion and usage of CO_2 in chemical reactions pose significant challenges, particularly owing to the stable nature of CO_2 [93].

In addition, carbonation curing is the procedure of subjecting hardened concrete to pure CO_2 and flue gas. CO_2 combines with different OPC phases and cement hydration products during the process, which is generally finished on the first day after mixing concrete [94]. This reaction creates stable minerals for sequestering carbon. The primary interactions of CO_2 with the OPC matrix are indicated in Equations 1–4.

$$3CaO.SiO_2 + 3CO_2 + xH_2O \rightarrow SiO_2.xH_2O + 3CaCO_3 \quad (1)$$

$$2CaO.SiO_2 + 2CO_2 + xH_2O \rightarrow SiO_2.xH_2O + 2CaCO_3 \quad (2)$$

$$Ca(OH)_2 + CO_2 \rightarrow H_2O + CaCO_3 \quad (3)$$

$$C-S-H + 2CO_2 \rightarrow SiO_2 + H_2O + 2CaCO_3 \quad (4)$$

The behavior of cement paste after being cured in the air for 18 h and subsequent exposure to CO_2 gas for 2 h was reported by Rostami et al. [95]. They discovered that specimens exposed to CO_2 produced higher hydration and carbonation products compared to specimens that were only exposed to air for 20 h. Furthermore, 8% of carbon in cement was captured by the carbonation process, resulting in a carbonation degree of 16%. This degree exhibits the amount of carbonated products developed compared to the binder's entire potential for carbonation. In mineral carbon capture (MCC) reactors, alkali wastes such as waste concrete and sludge are used to capture and store CO_2 as recarbonate that would have otherwise been released into the atmosphere. An illustration of CO_2 elimination from an OPC kiln is depicted in Fig. 10. Using various kinds of wastes, the MCC technology also makes CO_2 reduction possible. The decrement in CO_2 release can be determined from the stored recarbonate and the molar mass ratio of CO_2 to carbonates.

The recarbonate circle in the MCC technology reduces CO_2 emissions by reusing recarbonate in the cement production process. Direct inclusion of alkali wastes into cement kilns is restricted because of impurities like silicon. However, indirect MCC technologies aid in removing impurities while improving cement performance compared to OPC, as illustrated in Fig. 11.

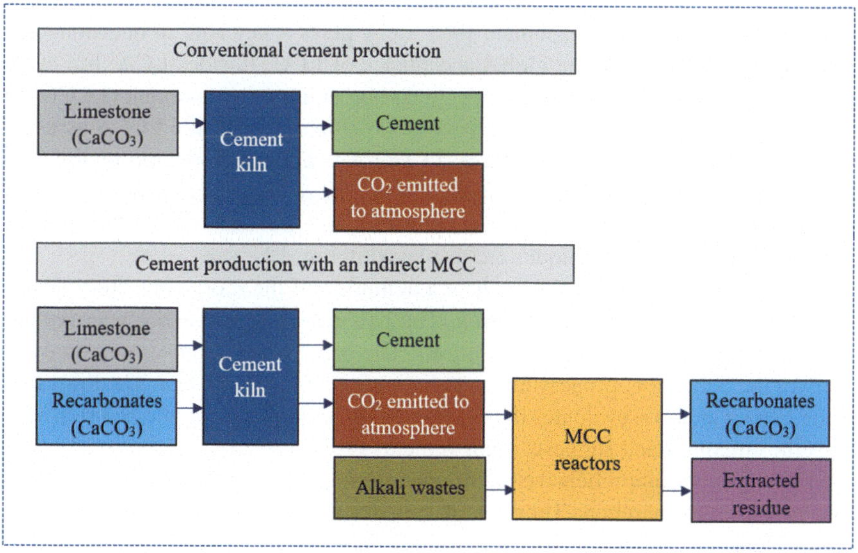

Fig. 10 Flowchart of CO_2 reduction from cement through carbon capture and utilization approach [90]

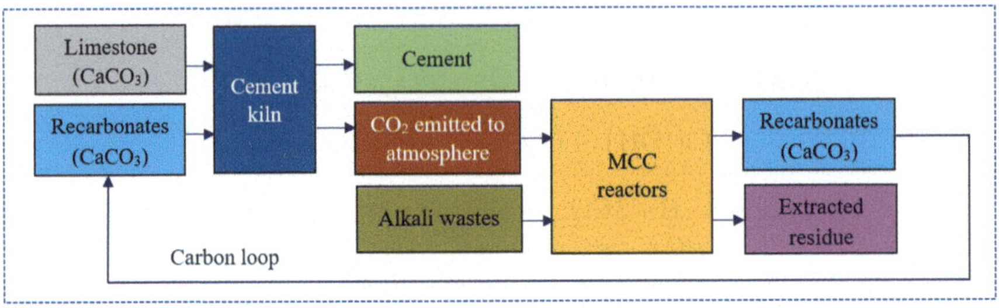

Fig. 11 Reutilization of recarbonates in cement production process [90]

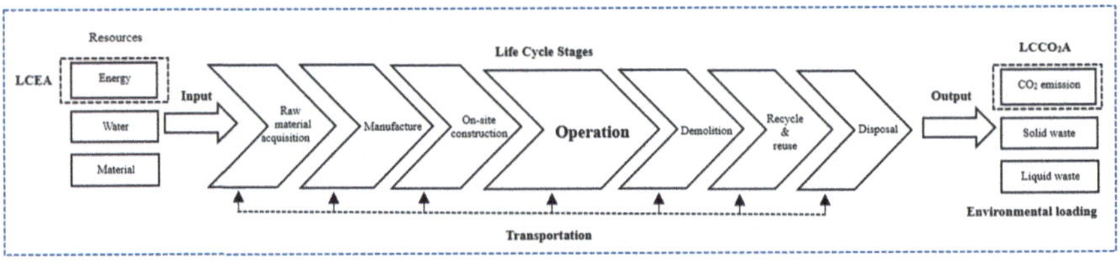

Fig. 12 Focus of LCEA, LCA, and $LCCO_2A$ [98]

4 Life Cycle and Environmental Impact Assessment of LCC

Life cycle assessment (LCA) is considered as an important factor in evaluating the sustainability of LCC. LCA helps point out key areas for improvement by considering the total lifetime of the materials. LCA not only identifies areas for improvement but also figures out the adverse effects of certain materials to avoid any negative impact on the environment [96]. LCA plays a key role in decision- and policymaking to improve the global acceptance of LCC. Besides LCA, life cycle energy assessment (LCEA) and life cycle carbon emission assessment ($LCCO_2A$) are also used to assess the overall impact of structures [97]. While LCEA assesses the energy utilization of the structures, $LCCO_2A$ is utilized to determine the CO_2 emission from a structure throughout its lifetime. Therefore, LCEA is not employed to determine the environmental impacts of a structure. Several steps of these aforementioned assessment techniques are displayed in Fig. 12.

Four important steps in LCA include a definition of the scope, inventory analysis, impact analysis, and interpretation. The boundaries, goals, and objectives are covered in the first step (scope definition). The second step of inventory analysis involves the collection, processing, and elimination of the input data. The third step of the impact analysis evaluates the environmental impacts, while the last step analyzes the outcomes and suggests improvements [98].

In addition, a comprehensive environmental impact analysis of LCC has been done by various researchers. Thorne et al. [99] determined the environmental impact of LCC, which incorporated SCMs. The study demonstrated that the introduction of limestone and FA has the potential to reduce carbon emissions, but it also decreases

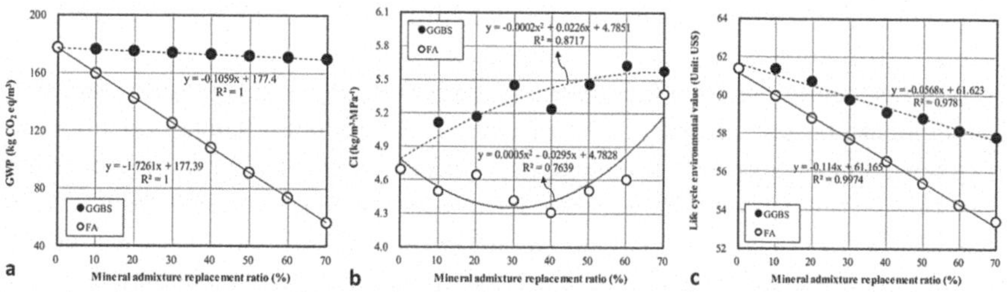

Fig. 13 Effect of FA and GGBS on: (**a**) GWP, (**b**) CO_2 intensity, and (**c**) LCA economic value [100]

the mechanical characteristics of concrete. Meanwhile, the OPC-FA mixture shows high strength and lower carbon emissions compared to the limestone–cement blend. Witte et al. [97] investigated the performance of LCC mixes and their influence on the environment through five LCA tools. The relationships between developed heat, porosity, and f'_c were assessed to resolve concerns regarding poor early-age strength and uncertain carbon accounting. The outcomes displayed that a high SCM mix could reduce GWP of concrete up to 50% (from 450 to 240 kg CO_2 eq/m^3), but variations were among various LCA tools. The research suggests an integrated strength eco-efficiency coefficient for the evaluation of concrete performance and influence on the environment simultaneously. Lee et al. [100] reported that increasing the replacement ratio of SCM resulted in the reduction of GWP by approximately 1.1 kg CO_2 eq/m^3 for GGBS and 17 kg CO_2 eq/m^3 for FA. Furthermore, the life cycle environmental impacts were considerably reduced too. The results indicated that an approximately 30% decrement in the CO_2 emission was achieved when f'_c of concrete was doubled. Additionally, another 15% reduction could be obtained with the SCM replacement, as illustrated in Fig. 13.

5 Challenges

Higher initial costs are among the top challenges in developing LCC for sustainable construction. For instance, the development of LCC involves carbon capture and utilization strategies, which require big investments. Also, the cost of SCMs such as GGBS is usually higher than the normal cement [101]. Besides the cost, every replacement material exhibits different behaviors under different conditions. Some of them are highly vulnerable to deterioration under extreme temperatures, posing a constraint for sustainable construction. Furthermore, getting rid of long-standing industry rules and opposition to change is mandatory for the global acceptance of sustainable building techniques. The reasons for this resistance include frequently ingrained conventions and professionals' unwillingness to adopt novel strategies or technical advancements. Challenges often arise from financial considerations, which emphasize upfront costs over long-term savings and environmental advantages. Complicating matters is a lack of knowledge about energy-efficient technologies and sustainable design, as well as problems with coordination throughout the construction supply chain. Comprehensive plans including training, public awareness schemes, teamwork, financial incentives, changes in legislation, and skill development startups are required to cater these problems. It takes an effort to spread awareness, knowledge, and support for change to transfer the construction industry's deeply ingrained habits toward a low-carbon and sustainable future. From a business point of view, investors are put under more uncertainty by unclear policy frameworks, which emphasize the need for open, encouraging policies. Investor confidence is hampered by the absence of trustworthy data about the financial returns and operational benefits of sustainable construction projects. Governments, business people, and financial professionals must work together to overcome these problems. Together, putting in place financial incentives, improving capital accessibility, encouraging knowledge sharing, and creating frameworks for supportive policies can push the transition to a built environment that is more resilient and sustainable.

6 Future Research Recommendations

Multiple researchers from all around the world have examined these LCCs. Nonetheless, there are still very few real-world uses for them in buildings [102]. It is possible to test the structural uses of the visually pleasing LCC flat slabs in a lab. Moreover, the effectiveness of load-bearing concrete walls composed of LCC has not been investigated comprehensively or used in building projects. While some glass fiber reinforced concretes (GFRC) have been utilized for building facades, other LCCs that incorporate plastic and rubber fragments have not been employed in building construction. Similar to this, industry and upcoming research partnerships will look into LCC-based prefabricated flooring, walls, and facades for residential and commercial structures.

Additionally, very limited research is available on the seismic, acoustic, and structural qualities of LCC for commercial applications. Research on the seismic performance of structures with reinforced LCC slabs, shear walls, columns, and beams that adhere to different structural guidelines is limited. Further research is also necessary to determine the bond strength, impact resistance, and dynamic loadings of LCC. LCC might work well for raft foundations, but this is still the frontier of its application. There is still research to be done on explosion simulations, fire resistance, and progressive collapse of LCC. Moreover, there are not enough comparative studies to determine which kind of LCC is best in terms of cost-effectiveness and structural suitability. Furthermore, to create hybrids and composites made of LCC for structural applications, further in-depth studies are required. Meanwhile, there has not been a detailed investigation of the thermal performance and energy efficiency of LCC buildings for use in residential and commercial settings.

Further research is important to evaluate how long-lasting LCC would be under harsh environmental conditions and corrosive surroundings. Thus, to assess performance under a variety of conditions, such as exposure to the sea and atmosphere, testing in both fast and real scenarios is required. Also, research is required to compare the durability of conventional LCC with three dimensional-printed LCC. Corrosion resistance, long-term performance in different conditions, and influence of chemical admixtures on f'_c are areas that need further research.

7 Conclusions

In conclusion, by decreasing carbon footprints and increasing resource utilization, LCC boosts eco-friendly building practices and sustainable construction. This chapter highlights the potential of concrete in production approaches, environmental impacts, and the utilization of new/alternative binders and technologies, laying out a roadmap for decreasing the negative influence of concrete production on the environment to resist climate change. The conclusions drawn from this chapter are as follows:

The use of SCMs in place of some OPC is a suitable method for lowering the carbon footprint of concrete while maintaining the durability and strength requirements. Potential clinker part substitutes include slag, CC, and limestone. This will reduce the CO_2 emissions that come with the cement manufacturing process. Like other SCMs, MK reduces CO_2 emissions while improving f'_c and durability of concrete when utilized as a half substitution for OPC. With less cement clinker content, CC, and limestone, LC3, a potential binder, dramatically lowers CO_2 emissions without sacrificing the qualities of concrete. Researchers compared f'_c and tensile strength of concrete employing CC, and it was noticed that concrete can attain equivalent or even higher strength values than an OPC system when utilizing up to 30% CC or a mixture of 50% CC and limestone in an LC3 system. By absorbing and reusing emitted CO_2, carbon capture and utilization technology contributes significantly to the mitigation of climate change by reducing greenhouse gases. In addition, carbon sequestration is facilitated by using captured CO_2 for LCC or other uses, which helps carbon-negative processes, including carbon mineralization for concrete manufacture. Moreover, LCA is a key process for thoroughly evaluating the environmental impact of concrete at every point of its life cycle, allowing stakeholders to make proper decisions. To tackle resistance to change within the complex network of the construction industry, joint efforts are required to address problems like market reluctance and material availability. In summary, LCC reduces CO_2 emissions, improves durability, and supports circular economy, making itself important for sustainable development and climate action. Its importance supports the objectives of global sustainability and emphasizes the necessity for more research into substitute materials and carbon capture methodologies. Its widespread acceptance will be fueled by the prioritization of life cycle studies and standardized certification systems, guaranteeing both sustainability and quality in construction procedures.

References

1. S.A. Miller, Supplementary cementitious materials to mitigate greenhouse gas emissions from concrete: Can there be too much of a good thing? J. Clean. Prod. **178**, 587–598 (2018). https://doi.org/10.1016/J.JCLEPRO.2018.01.008
2. M.Ö. Arıoğlu Akan, D.G. Dhavale, J. Sarkis, Greenhouse gas emissions in the construction industry: An analysis and evaluation of a concrete supply chain. J. Clean. Prod. **167**, 1195–1207 (2017). https://doi.org/10.1016/J.JCLEPRO.2017.07.225
3. E. Benhelal, G. Zahedi, E. Shamsaei, A. Bahadori, Global strategies and potentials to curb CO2 emissions in cement industry. J. Clean. Prod. **51**, 142–161 (2013). https://doi.org/10.1016/J.JCLEPRO.2012.10.049
4. D.R. Nhuchhen, S.P. Sit, D.B. Layzell, Decarbonization of cement production in a hydrogen economy. Appl. Energy **317**, 119180 (2022). https://doi.org/10.1016/J.APENERGY.2022.119180
5. Q.D. Nguyen, M.S.H. Khan, A. Castel, T. Kim, Durability and microstructure properties of low-carbon concrete incorporating ferronickel slag sand and fly ash. J. Mater. Civ. Eng. **31**(8), 04019152 (2019). https://doi.org/10.1061/(ASCE)MT.1943-5533.0002797
6. P.D. Nukah, S.J. Abbey, C.A. Booth, J. Oti, Development of low carbon concrete and prospective of geopolymer concrete using lightweight coarse aggregate and cement replacement materials. Constr. Build. Mater. **428**, 136295 (2024). https://doi.org/10.1016/J.CONBUILDMAT.2024.136295
7. K.L. Scrivener, V.M. John, E.M. Gartner, Eco-efficient cement: Potential economically viable solutions for a low-CO_2 cement-based materials industry. Cem. Concr. Res. **114**, 2–26 (2018). https://doi.org/10.1016/J.CEMCONRES.2018.03.015
8. H. Guo et al., Durability of recycled aggregate concrete—a review. Cem. Concr. Compos. **89**, 251–259 (2018). https://doi.org/10.1016/J.CEMCONCOMP.2018.03.008
9. N. Zhang, H. Duan, T.R. Miller, V.W.Y. Tam, G. Liu, J. Zuo, Mitigation of carbon dioxide by accelerated sequestration in concrete debris. Renew. Sust. Energ. Rev. **117**, 109495 (2020). https://doi.org/10.1016/J.RSER.2019.109495
10. L.M. Haselbach, J.N. Thomle, An alternative mechanism for accelerated carbon sequestration in concrete. Sustain. Cities Soc. **12**, 25–30 (2014). https://doi.org/10.1016/J.SCS.2014.01.001
11. Z. Liu, W. Meng, Fundamental understanding of carbonation curing and durability of carbonation-cured cement-based composites: A review. J. CO2 Util. **44**, 101428 (2021). https://doi.org/10.1016/J.JCOU.2020.101428
12. J. Skocek, M. Zajac, M.B. Haha, Carbon capture and utilization by mineralization of cement pastes derived from recycled concrete. Sci Rep **10**(1), 1–12 (2020). https://doi.org/10.1038/s41598-020-62503-z
13. H.F. Campos, N.S. Klein, J. Marques Filho, Proposed mix design method for sustainable high-strength concrete using particle packing optimization. J. Clean. Prod. **265**, 121907 (2020). https://doi.org/10.1016/J.JCLEPRO.2020.121907
14. V. Shobeiri, B. Bennett, T. Xie, P. Visintin, Mix design optimization of concrete containing fly ash and slag for global warming potential and cost reduction. Case Stud. Constr. Mater. **18**, e01832 (2023). https://doi.org/10.1016/J.CSCM.2023.E01832
15. V. Sirivivatnanon, C. Xue, R. Khatri, Long-term reinforcement corrosion in low carbon concrete with a high volume of SCMs exposed to NaCl solutions and field marine environment. Constr. Build. Mater. **393**, 132071 (2023). https://doi.org/10.1016/J.CONBUILDMAT.2023.132071
16. C.M. Martín, N.B. Scarponi, Y.A. Villagrán, D.G. Manzanal, T.M. Piqué, Pozzolanic activity quantification of hollow glass microspheres. Cem. Concr. Compos. **118**, 103981 (2021). https://doi.org/10.1016/J.CEMCONCOMP.2021.103981
17. J. Duchesne, Alternative supplementary cementitious materials for sustainable concrete structures: A review on characterization and properties. Waste Biomass Valori. **12**(3), 1219–1236 (2021). https://doi.org/10.1007/S12649-020-01068-4/FIGURES/3

18. S. Gupta, S. Chaudhary, State of the art review on supplementary cementitious materials in India – II: Characteristics of SCMs, effect on concrete and environmental impact. J. Clean. Prod. **357**, 131945 (2022). https://doi.org/10.1016/J.JCLEPRO.2022.131945
19. B.G. Kutchko, A.G. Kim, Fly ash characterization by SEM–EDS. Fuel **85**(17–18), 2537–2544 (2006). https://doi.org/10.1016/J.FUEL.2006.05.016
20. F. Wang, X. Sun, Z. Tao, Z. Pan, Effect of silica fume on compressive strength of ultra-high-performance concrete made of calcium aluminate cement/fly ash based geopolymer. J. Build. Eng. **62**, 105398 (2022). https://doi.org/10.1016/J.JOBE.2022.105398
21. Y. Zhu et al., Effect of decarbonization of high carbon fly ash on workability, mechanical properties and durability of concrete. Mater. Struct./Materiaux et Constructions **56**(9), 1–15 (2023). https://doi.org/10.1617/S11527-023-02258-X/FIGURES/11
22. M.H. Kumar, N.R. Mohanta, N. Patel, S. Samantaray, S.V.B. Reddy, Impact of fly ash and metakaoline on the crack resistance and shrinkage of concrete. Iran. J. Sci. Technol. Transact. Civil Eng. **46**(3), 2011–2026 (2022). https://doi.org/10.1007/S40996-021-00748-Y/FIGURES/12
23. E. Grabias-Blicharz, W. Franus, A critical review on mechanochemical processing of fly ash and fly ash-derived materials. Sci. Total Environ. **860**, 160529 (2023). https://doi.org/10.1016/J.SCITOTENV.2022.160529
24. M. Amran et al., Fly ash-based eco-efficient concretes: A comprehensive review of the short-term properties. Materials **14**(15), 4264 (2021). https://doi.org/10.3390/MA14154264
25. C. Herath, C. Gunasekara, D.W. Law, S. Setunge, Long term mechanical performance of nano-engineered high volume fly ash concrete. J. Build. Eng. **43**, 103168 (2021). https://doi.org/10.1016/J.JOBE.2021.103168
26. G.R. Woolley, R.M. Conlin, M.R.E. Joyce, Discussion. Pulverised fuel ash concrete in construction of natural draught cooling towers. **86**(6), 1205–1207 (2015). https://doi.org/10.1680/iicep.1989.3663
27. G. Li, C. Zhou, W. Ahmad, K.I. Usanova, M. Karelina, A.M. Mohamed, R. Khallaf, Fly ash application as supplementary cementitious material: a review. Materials ***15***(7), 2664 (2022). https://doi.org/10.3390/MA15072664
28. Y. Hou, D. Wang, W. Zhou, H. Lu, L. Wang, Study on sulfate-resistance of fly ash-based geopolymers. New Build. Mater. **7**, 41–44 (2008)
29. A. Palomo, M.W. Grutzeck, M.T. Blanco, Alkali-activated fly ashes: A cement for the future. Cem. Concr. Res. **29**(8), 1323–1329 (1999). https://doi.org/10.1016/S0008-8846(98)00243-9
30. S. Thokchom, P. Ghosh, S. Ghosh, Performance of fly ash based geopolymer mortars in sulphate solution. J. Eng. Sci. Technol. Rev. **3**(1), 36–40 (2010)
31. M. Barbuta, R. Bucur, A.A. Serbanoiu, S. Scutarasu, A. Burlacu, Combined effect of fly ash and fibers on properties of cement concrete. Procedia Eng. **181**, 280–284 (2017). https://doi.org/10.1016/J.PROENG.2017.02.390
32. O. Gencel, F. Koksal, C. Ozel, W. Brostow, Combined effects of fly ash and waste ferrochromium on properties of concrete. Constr. Build. Mater. **29**, 633–640 (2012). https://doi.org/10.1016/J.CONBUILDMAT.2011.11.026
33. C.H. Huang, S.K. Lin, C.S. Chang, H.J. Chen, Mix proportions and mechanical properties of concrete containing very high-volume of class F fly ash. Constr. Build. Mater. **46**, 71–78 (2013). https://doi.org/10.1016/J.CONBUILDMAT.2013.04.016
34. M. Şahmaran, S.B. Keskin, G. Ozerkan, I.O. Yaman, Self-healing of mechanically-loaded self consolidating concretes with high volumes of fly ash. Cem. Concr. Compos. **30**(10), 872–879 (2008). https://doi.org/10.1016/J.CEMCONCOMP.2008.07.001
35. K. Celik, C. Meral, M. Mancio, P.K. Mehta, P.J.M. Monteiro, A comparative study of self-consolidating concretes incorporating high-volume natural pozzolan or high-volume fly ash. Constr. Build. Mater. **67**, 14–19 (2014). https://doi.org/10.1016/J.CONBUILDMAT.2013.11.065

36. M. Şahmaran, I.Ö. Yaman, M. Tokyay, Transport and mechanical properties of self consolidating concrete with high volume fly ash. Cem. Concr. Compos. **31**(2), 99–106 (2009). https://doi.org/10.1016/J.CEMCONCOMP.2008.12.003
37. A. Mehta, D.K. Ashish, Silica fume and waste glass in cement concrete production: A review. J. Build. Eng. **29**, 100888 (2020). https://doi.org/10.1016/J.JOBE.2019.100888
38. B.W. Jo, C.H. Kim, G.H. Tae, J.B. Park, Characteristics of cement mortar with nano-SiO2 particles. Constr. Build. Mater. **21**(6), 1351–1355 (2007). https://doi.org/10.1016/J.CONBUILDMAT.2005.12.020
39. R. Suryanita, H. Maizir, R. Zulapriansyah, Y. Subagiono, M.F. Arshad, The effect of silica fume admixture on the compressive strength of the cellular lightweight concrete. Results Eng. **14**, 100445 (2022)
40. H.M. Hamada et al., Effect of silica fume on the properties of sustainable cement concrete. J. Mater. Res. Technol. **24**, 8887–8908 (2023)
41. M. Gupta, R. Raj, A.K. Sahu, Effect of rice husk ash, silica fume & GGBFS on compressive strength of performance based concrete. Mater. Today Proc. **55**, 234–239 (2022). https://doi.org/10.1016/J.MATPR.2021.06.343
42. M. Abdi Moghadam, R.A. Izadifard, Experimental investigation on the effect of silica fume and zeolite on mechanical and durability properties of concrete at high temperatures. SN Appl. Sci. **1**(7), 1–11 (2019). https://doi.org/10.1007/S42452-019-0739-2/FIGURES/14
43. S. Tak, P. Gupta, A. Kumar, A. Sofi, C. Mei Yun, Effect of using silica fume as a partial replacement of cement in concrete. Mater. Today Proc. (2023). https://doi.org/10.1016/J.MATPR.2023.04.205
44. M.I. Khan, R. Siddique, Utilization of silica fume in concrete: Review of durability properties. Resour. Conserv. Recycl. **57**, 30–35 (2011). https://doi.org/10.1016/J.RESCONREC.2011.09.016
45. A.K. Chishi, L. Gautam, Sustainable use of silica fume in green cement concrete production: A review. Innov. Infrastruct. Solut. **8**(7), 1–14 (2023). https://doi.org/10.1007/S41062-023-01164-Z/FIGURES/5
46. F. Alberto, Z. Sommariva, Limestone calcined clay cements (LC3): Raw material processing, sulfate balance and hydration kinetics, (2020). https://doi.org/10.5075/EPFL-THESIS-8173.
47. S. Barbhuiya, J. Nepal, B.B. Das, Properties, compatibility, environmental benefits and future directions of limestone calcined clay cement (LC3) concrete: A review. J. Build. Eng. **79**, 107794 (2023). https://doi.org/10.1016/J.JOBE.2023.107794
48. S.A. Joseph, Y. Dhandapani, D. Geddes, Z. Zhao, S. Bishnoi, M. Vieira, F. Martirena, A. Castel, F. Kanavaris, T. Bansal, K.A. Riding, et al., Mechanical properties of concrete made with calcined clay: a review by RILEM TC-282 CCL. Mater. Struct. **56**(4), 1–20 (2023). https://doi.org/10.1617/S11527-023-02118-8
49. B. Sabir, S. Wild, J. Bai, Metakaolin and calcined clays as pozzolans for concrete: A review. Cem. Concr. Compos. **23**(6), 441–454 (2001). https://doi.org/10.1016/S0958-9465(00)00092-5
50. Y. Dhandapani, K. Vignesh, T. Raja, M. Santhanam, Development of the microstructure in LC3 systems and its effect on concrete properties. RILEM Bookseries **16**, 131–140 (2018). https://doi.org/10.1007/978-94-024-1207-9_21
51. Y.R. Alharbi, A.A. Abadel, A.A. Salah, O.A. Mayhoub, M. Kohail, Engineering properties of alkali activated materials reactive powder concrete. Constr. Build. Mater. **271**, 121550 (2021). https://doi.org/10.1016/J.CONBUILDMAT.2020.121550
52. A. Wang et al., The durability of alkali-activated materials in comparison with ordinary Portland cements and concretes: A review. Engineering **6**(6), 695–706 (2020). https://doi.org/10.1016/J.ENG.2019.08.019
53. J. Davidovits, Geopolymers and geopolymeric materials. J. Therm. Anal. **35**(2), 429–441 (1989). https://doi.org/10.1007/BF01904446/METRICS
54. J. Davidovits, Geopolymers of the first generation: SILIFACE-process. Geopolymer, 49–67 (1988)

55. H.A. Alcamand, P.H.R. Borges, F.A. Silva, A.C.C. Trindade, The effect of matrix composition and calcium content on the sulfate durability of metakaolin and metakaolin/slag alkali-activated mortars. *Ceram. Int.* **44**(5), 5037–5044 (2018). https://doi.org/10.1016/J.CERAMINT.2017.12.102
56. X. Yan, L. Jiang, M. Guo, Y. Chen, Z. Song, R. Bian, Evaluation of sulfate resistance of slag contained concrete under steam curing. Constr. Build. Mater. **195**, 231–237 (2019). https://doi.org/10.1016/J.CONBUILDMAT.2018.11.073
57. W.H. Tao, X.H. Fu, F.J. Sun, Z.X. Yang, Studies on properties and mechanisms of geopolymer cementitious material. Bull. Chin. Ceram. Soc. **27**, 730–739 (2008)
58. M. Valizadeh Kiamahalleh, A. Gholampour, Y. Tang, T.D. Ngo, Incorporation of reduced graphene oxide in waste-based concrete including lead smelter slag and recycled coarse aggregate. J. Build. Eng. **88**, 109221 (2024). https://doi.org/10.1016/J.JOBE.2024.109221
59. I.F. Sáez del Bosque, P. Van den Heede, N. De Belie, M.I. Sánchez de Rojas, C. Medina, Carbonation of concrete with construction and demolition waste based recycled aggregates and cement with recycled content. Constr. Build. Mater. **234**, 117336 (2020). https://doi.org/10.1016/J.CONBUILDMAT.2019.117336
60. J. Acordi, A. Luza, D.C.N. Fabris, F. Raupp-Pereira, A. De Noni, O.R.K. Montedo, New waste-based supplementary cementitious materials: Mortars and concrete formulations. Constr. Build. Mater. **240**, 117877 (2020). https://doi.org/10.1016/J.CONBUILDMAT.2019.117877
61. M. Contreras Llanes, M. Romero Pérez, M.J. Gázquez González, J.P. Bolívar Raya, Construction and demolition waste as recycled aggregate for environmentally friendly concrete paving. Environ. Sci. Pollut. Res. **29**(7), 9826–9840 (2022). https://doi.org/10.1007/S11356-021-15849-4/TABLES/6
62. P.H. Dalal, M. Patil, K.K.R. Iyer, T.N. Dave, Sustainable controlled low strength material from waste materials for infrastructure applications: State-of-the-art. J. Environ. Manag. **342**, 118284 (2023). https://doi.org/10.1016/J.JENVMAN.2023.118284
63. S.N. Shah, K.H. Mo, S.P. Yap, J. Yang, T.C. Ling, Lightweight foamed concrete as a promising avenue for incorporating waste materials: A review. Resour. Conserv. Recycl. **164**, 105103 (2021). https://doi.org/10.1016/J.RESCONREC.2020.105103
64. K. McNeil, T.H.K. Kang, Recycled concrete aggregates: A review. Int. J. Concr. Struct. Mater. **7**(1), 61–69 (2013). https://doi.org/10.1007/S40069-013-0032-5/FIGURES/6
65. J. Wang, J. Zhang, D. Cao, H. Dang, B. Ding, Comparison of recycled aggregate treatment methods on the performance for recycled concrete. Constr. Build. Mater. **234**, 117366 (2020). https://doi.org/10.1016/J.CONBUILDMAT.2019.117366
66. A.S. Alqarni, H. Abbas, K.M. Al-Shwikh, Y.A. Al-Salloum, Treatment of recycled concrete aggregate to enhance concrete performance. Constr. Build. Mater. **307**, 124960 (2021)
67. J. Feng, H. Zhang, K. Gao, Y. Liao, W. Gao, G. Wu, Efficient creep prediction of recycled aggregate concrete via machine learning algorithms. Constr. Build. Mater. **360**, 129497 (2022)
68. X. Yuan et al., Machine learning prediction models to evaluate the strength of recycled aggregate concrete. Materials **15**(8), 2823 (2022)
69. K. Liu, J. Zheng, S. Dong, W. Xie, X. Zhang, Mixture optimization of mechanical, economical, and environmental objectives for sustainable recycled aggregate concrete based on machine learning and metaheuristic algorithms. J. Build. Eng. **63**, 105570 (2023)
70. Y. Peng, C. Unluer, Modeling the mechanical properties of recycled aggregate concrete using hybrid machine learning algorithms. Resour. Conserv. Recycl. **190**, 106812 (2023). https://doi.org/10.1016/J.RESCONREC.2022.106812
71. T.E. Frank, M.D. Lepech, S.L. Billington, Experimental testing of reinforced ECC beams subjected to various cyclic deformation histories. J. Struct. Eng. **144**(6), 04018052 (2018). https://doi.org/10.1061/(ASCE)ST.1943-541X.0002034
72. Y. Agrawal, T. Gupta, R. Sharma, N.L. Panwar, S. Siddique, A comprehensive review on the performance of structural lightweight aggregate concrete for sustainable construction. Constr. Mater. **1**(1), 39–62 (2021)

73. F. Hussain, F. Rehman, R.A. Khushnood, S. Ali Khan, A. Hamza, Study of physical and mechanical behavior of artificial lightweight aggregate made of Pakistani clays. Constr. Build. Mater. **378**, 131103 (2023). https://doi.org/10.1016/J.CONBUILDMAT.2023.131103
74. F. Hussain, A. Omran, N. Soliman, Synthesis and characterization of bentonite-based lightweight ceramic aggregates using coal combustion residue and kerosene bloating agent. Constr. Build. Mater. **425**, 135916 (2024)
75. Z. Liu et al., Carbon monitor, a near-real-time daily dataset of global CO_2 emission from fossil fuel and cement production. Sci. Data **7**(1), 392 (2020)
76. M.M.H. Khan, J. Havukainen, M. Horttanainen, Impact of utilizing solid recovered fuel on the global warming potential of cement production and waste management system: A life cycle assessment approach. Waste Manag. Res. **39**(4), 561–572 (2021)
77. T. Joensuu, R. Leino, J. Heinonen, A. Saari, Developing buildings' life cycle assessment in circular economy-comparing methods for assessing carbon footprint of reusable components. Sustain. Cities Soc. **77**, 103499 (2022)
78. N. Mohamad, K. Muthusamy, R. Embong, A. Kusbiantoro, M.H. Hashim, Environmental impact of cement production and solutions: A review. Mater. Today Proc. **48**, 741–746 (2022). https://doi.org/10.1016/J.MATPR.2021.02.212
79. G.S. Kishan, M. Sakthivel, R. Vijayakumar, N. Lingeshwaran, Life cycle assessment on tire derived fuel as alternative fuel in cement industry. Mater. Today Proc. **47**, 5483–5488 (2021)
80. K. Wojtacha-Rychter, P. Kucharski, A. Smolinski, Conventional and alternative sources of thermal energy in the production of cement—An impact on CO_2 emission. Energies (Basel) **14**(6), 1539 (2021)
81. A.L.H.R. El-Salamony, H.M. Mahmoud, N. Shehata, Enhancing the efficiency of a cement plant kiln using modified alternative fuel. Environ. Nanotechnol. Monit. Manag. **14**, 100310 (2020). https://doi.org/10.1016/J.ENMM.2020.100310
82. A. Adesina, P. Awoyera, Utilization of biomass energy in cement production: A pathway towards sustainable infrastructure, pp. 791–799, 2020, doi: https://doi.org/10.1007/978-3-030-18488-9_65
83. G. Asadollahfardi, M. Delnavaz, V. Rashnoiee, N. Ghonabadi, Use of treated domestic wastewater before chlorination to produce and cure concrete. Constr. Build. Mater. **105**, 253–261 (2016). https://doi.org/10.1016/J.CONBUILDMAT.2015.12.039
84. Y. Liu et al., Utilization of drinking water treatment sludge in concrete paving blocks: Microstructural analysis, durability and leaching properties. J. Environ. Manag. **262**, 110352 (2020). https://doi.org/10.1016/J.JENVMAN.2020.110352
85. H. Yoonus, S.G. Al-Ghamdi, Environmental performance of building integrated grey water reuse systems based on life-cycle assessment: A systematic and bibliographic analysis. Sci. Total Environ. **712** (2020). https://doi.org/10.1016/J.SCITOTENV.2020.136535
86. H. Varshney, R.A. Khan, I.K. Khan, Sustainable use of different wastewater in concrete construction: A review. J. Build. Eng. ***41***, 102411 (2021). https://doi.org/10.1016/J.JOBE.2021.102411
87. S. Mane, S. Faizal, G. Prakash, S. Bhandarkar, V. Kumar, Use of sewage treated water in concrete. Int. J. Res. Eng. Sci. Manag. **2**(6), 210–213 (2019)
88. M.N.N. Khan, A.K. Saha, P.K. Sarker, Reuse of waste glass as a supplementary binder and aggregate for sustainable cement-based construction materials: A review. J. Build. Eng. **28**, 101052 (2020)
89. U.J. Malik et al., ANN-based predictive mimicker for the constitutive model of engineered cementitious composites (ECC). Constr. Build. Mater. **420**, 135530 (2024). https://doi.org/10.1016/J.CONBUILDMAT.2024.135530
90. Y. Izumi, A. Iizuka, H.J. Ho, Calculation of greenhouse gas emissions for a carbon recycling system using mineral carbon capture and utilization technology in the cement industry. J. Clean. Prod. **312**, 127618 (2021). https://doi.org/10.1016/J.JCLEPRO.2021.127618
91. V. Srinivasan et al., *CO_2 utilisation roadmap* (CSIRO Australia, 2021)

92. D. Meng, C. Unluer, E.-H. Yang, S. Qian, Carbon sequestration and utilization in cement-based materials and potential impacts on durability of structural concrete. Constr. Build. Mater. **361**, 129610 (2022)
93. A. Al-Mamoori, A. Krishnamurthy, A.A. Rownaghi, F. Rezaei, Carbon capture and utilization update. Energ. Technol. **5**(6), 834–849 (2017). https://doi.org/10.1002/ENTE.201600747
94. M. Kazemian, B. Shafei, Carbon sequestration and storage in concrete: A state-of-the-art review of compositions, methods, and developments. J. CO2 Util. **70**, 102443 (2023)
95. V. Rostami, Y. Shao, A.J. Boyd, Z. He, Microstructure of cement paste subject to early carbonation curing. Cem. Concr. Res. **42**(1), 186–193 (2012)
96. S. Chen, Y. Teng, Y. Zhang, C.K.Y. Leung, W. Pan, Reducing embodied carbon in concrete materials: A state-of-the-art review. Resour. Conserv. Recycl. **188**, 106653 (2023)
97. A. Witte, N. Garg, Quantifying the global warming potential of low carbon concrete mixes: Comparison of existing life cycle analysis tools. Case Stud. Constr. Mater. **20**, e02832 (2024)
98. C.K. Chau, T.M. Leung, W.Y. Ng, A review on life cycle assessment, life cycle energy assessment and life cycle carbon emissions assessment on buildings. Appl. Energy **143**, 395–413 (2015)
99. J. Thorne, D.V. Bompa, M.F. Funari, N. Garcia-Troncoso, Environmental impact evaluation of low-carbon concrete incorporating fly ash and limestone. Clean. Mater. **12**, 100242 (2024). https://doi.org/10.1016/J.CLEMA.2024.100242
100. J. Lee, T. Lee, J. Jeong, J. Jeong, Sustainability and performance assessment of binary blended low-carbon concrete using supplementary cementitious materials. J. Clean. Prod. **280**, 124373 (2021). https://doi.org/10.1016/J.JCLEPRO.2020.124373
101. C. Zhang et al., Environmental life cycle costing at the early stage for supporting cost optimization of precast concrete panel for energy renovation of existing buildings. J. Build. Eng. **35**, 102002 (2021)
102. M. Wasim, A. Abadel, B.H. Abu Bakar, I.M.H. Alshaikh, Future directions for the application of zero carbon concrete in civil engineering—A review. Case Stud. Constr. Mater. **17**, e01318 (2022). https://doi.org/10.1016/J.CSCM.2022.E01318

Developing Empirical Models for Assessing On-Site Performance of Structural Concrete Elements through Nondestructive Testing Methods

Kolandasamy Poongodi, Alireza Bahrami, Palanisamy Murthi, and Krishnasamy Thirumalairaja

1 Introduction

Construction quality assessment is an important task for field engineers. The complete achievement of any project depends on the material testing at the laboratory and quality evaluation at site. Both material testing and onsite inspection are essential in a construction project to ensure the functional performance and service life of the project. Apart from the new construction, periodical investigations into an existing structure are mandatory as a matter of health monitoring of structures. Though various on-site testing can be adopted to evaluate the performance of structural elements, a few tests are essential to prove the performance of the structural elements. In reinforced cement concrete (RCC) structures, concrete strength plays an essential part in ensuring compliance with the requirements of design parameters and codal requirements. The strength tests are normally done using specimens prepared in the laboratory environment and collection of samples from the construction site during execution. The on-site evaluation of the strength properties of structural components is much needed to ensure the functional performance of the structures [1]. The

K. Poongodi (✉)
Department of Civil Engineering, SR University, Warangal, India
e-mail: k.poongodi@sru.edu.in

A. Bahrami (✉)
Department of Building Engineering, Energy Systems and Sustainability Science, Faculty of Engineering and Sustainable Development, University of Gävle, Gävle, Sweden
e-mail: alireza.bahrami@hig.se

P. Murthi
Department of General Engineering, Cheran College of Technology, Kangeyam, India

K. Thirumalairaja
Department of Civil Engineering, SNS College of Technology, Coimbatore, India

A. Bahrami (ed.), *Advances in Sustainable Concrete for Construction*,
https://doi.org/10.1007/978-3-031-85052-3_4

destructive method of on-site testing to predict the strength properties of a completed portion of structural components is not feasible, and a nondestructive approach is the only possible way.

These nondestructive approaches have been growing in popularity to measure the quality of concrete due to their simplicity and quick assessment. In recent years, the nondestructive approach has proven to be an adoptable method for quality assessment of concrete in an existing concrete structure. Various nondestructive techniques are being practiced to check the performance of structures. However, Schmidt rebound hammer (SRH) and ultrasonic pulse velocity (UPV) tests have been extensively utilized as nondestructive approaches. These two methods are very effective and easy to execute. Hence, these two methods have been used to estimate concrete strength properties for the last few decades [2, 3]. It was reported that the effectiveness of these two methods depends on the void condition, age, and humidity of concrete [4].

In a UPV test, ultrasonic waves are generated by a high voltage pulse. The waves are transmitted to concrete through one transducer and collected by the other transducer placed at the other end. The time it takes to pass the ultrasonic wave through concrete primarily depends on its density and elasticity. From the time taken to transmit the wave, the pulse velocity can be calculated based on the distance travelled through the concrete mass. The results from the UPV test can be correlated to the concrete performance qualitatively such as excellent quality (the pulse velocity (V) more than 4.5 km/s), good ($V = 3.5$ to 4.5 km/s), medium ($V = 3$ to 3.4 km/s), and doubtful quality ($V < 3$ km/s) according to IS: 13311–1992 (Part-1) [5].

On the other hand, SRH is used for assessing the surface hardness of concrete and expressed as rebound hammer number (RN). From RN, the compressive strength (f_c) can be correlated by way of a chart affixed to the device after standard calibration. It is possible to evaluate the surface hardness up to 30 to 50 mm below the surface through the SRH test [6]. Besides, the supplied chart does not reflect the confidence results because of continuous usage and testing of concrete with different compositions due to the latest development in concrete technology. The SRH test is used to detect homogeneities and assess the quality of concrete. The SRH tests have been carried out with various precautionary procedures, like the test surface should be free from any scaling, honeycombing, rough texture, cracks, or porous nature. The moisture condition should be uniform throughout the surface [7].

The strength properties of concrete in an existing structure had been predicted with reference to the experimental relationship between the actual f_c found in a laboratory investigation and the nondestructive test (NDT) results obtained from the same specimens [8]. It is difficult to assess the mechanical properties of concrete using NDT methods directly. The empirical models are more evident in evaluating the mechanical properties. For SRH, the relationship chart is affixed to the device, which reflects the experimental investigation conducted by the manufacturer using their system of testing. The prescribed correlation may not offer the genuine strength performance of a real structure. The number of variable parameters affects the prediction of strength properties using NDT evaluation, such as differences in concrete type and grade, differences in mix proportioning, selection of aggregates, and w/c ratio [9]. With the aim of predicting reliable and accurate results for achieving the compressive strength of on-site concrete, the NDT results are correlated with the strength obtained from the core specimens (semi-NDT test), which are extracted from the cast-in-situ concrete elements [10–12]. In order to overcome the difficulties, an appropriate correlation between RN and f_c of concrete has to be developed for determining the strength performance of existing structures. To validate the actual strength properties of an already built structure, it is essential to establish an acceptable correlation between the NDT evaluation and f_c. In this background, the study aimed at developing the most appropriate correlation for conventional concrete from normal strength to high strength and pumpable concrete, which is suitable for ready-mixed concrete (RMC).

2 Investigation Program

2.1 *Methodology*

2.1.1 UPV

The quality of concrete is predicted by doing UPV test as per the guidelines of IS: 13311-1992 (Part-1) [5]. Figures 1 and 2 show the UPV test performed in a laboratory environment and on-site, respectively. The concept of the UPV method is based on the travelling time of the waves from the emitter transducer to the receiver transducer, and the pulse velocity can be calculated as:

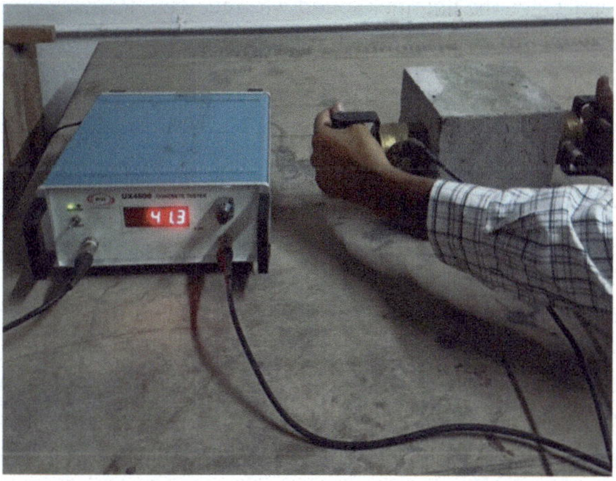

Fig. 1 UPV test at laboratory

Fig. 2 UPV test at site

$$V = \frac{L}{t} \tag{1}$$

where V = pulse velocity (km/s), L = path length (mm), and t = travelling time (μs).

2.1.2 RN

The SRH test was carried out with reference to IS: 13311-1992 (Part-2) [13]. Selecting a uniform surface for the SRH test was the first step, and the surface should be smoothed before conducting the test. The plunger of SRH was positioned perpendicular to the smooth surface, and the hammer was pushed on the test surface. Once the hammer sound was received, RN was carefully recorded on the scale. Figures 3 and 4 display the SRH test conducted in a laboratory environment and on-site, respectively.

2.2 Materials

The cast-in-situ concrete cubes of size 150 mm × 150 mm × 150 mm were cast using ordinary Portland cement (OPC) according to IS: 269-2015 [14] in the case I study. The tested values of OPC specifications are listed in Table 1. Crushed granite coarse aggregate (CA) of sizes 10 mm and 20 mm and natural river sand belonging to grade zone II were selected for casting the concrete cubes. Table 2 presents the physical properties of aggregates. In case II, the concrete cubes were cast at delivery point of RMC at various sites. The cement and aggregate samples were collected from the respective RMC plants for testing as per IS: 383-1970 [15], and the results

Fig. 3 SRH test at laboratory

Fig. 4 SRH tests at site

are depicted in Table 3. The grading zone curves for the selected aggregates are revealed in Fig. 5. In total, 75 concrete cubes were cast, and the mean values of three cubes were considered in this investigation considering 3 days, 7 days, 14 days, 28 days, and 90 days of curing.

Table 1 OPC specifications

Chemical composition (%)	CaO	SiO_2	Fe_2O_3	Al_2O_3	Na_2O	SO_3	MgO	LoI
	67.64	18.35	3.21	4.27	0.17	2.14	0.89	1.15
Physical properties	Specific gravity		Initial setting time		Final setting time		Compressive strength (28 days)	
	3.15		75 minutes		165 minutes		54.37 MPa	

Table 2 Properties of aggregates used in cast-in-situ laboratory investigation

Properties	Fine aggregate	10 mm coarse aggregate	20 mm coarse aggregate
Specific gravity	2.57	2.63	2.79
Grading zone	II	Single sized	Single sized

Table 3 Properties of ingredients utilized for pumpable concrete

Properties	Cement	Fine aggregate	10 mm coarse aggregate	20 mm coarse aggregate
Specific gravity	3.15	2.61	2.58	2.65
Grading zone	–	II	Single sized	Single sized
Compressive strength (28 days)	53.75 MPa	–	–	–

2.3 Mix Proportioning

The normal concrete of M20 grade to the high-strength concrete of M60 grade was developed for cast-in-situ concrete to evaluate the nondestructive parameters using the UPV test and the SRH test in the case I study in a laboratory environment. The quantities of ingredients for 1 m^3 of the selected grades are indicated in Table 4. The mix proportioning of various pumpable grades of concrete was sourced from the respective units and mentioned in Table 5. Low-calcium fly ash (class F type) was selected for developing the mix proportioning of M40-, M50-, and M60-grade concretes and sulfonated naphthalene polymer-based superplasticizer conforming to ASTM C-494 [16].

3 Results and Discussion

3.1 Correlation Methods

Three different approaches have been examined by researchers to predict the most reliable relationship between f_c and NDT results, such as linear, exponential, and power correlation methods. The general forms of these three correlation methods are listed below:

Fig. 5 Grade zone curves for selected aggregates: (**a**) 10 mm coarse aggregate, (**b**) 20 mm coarse aggregate, and (**c**) river sand

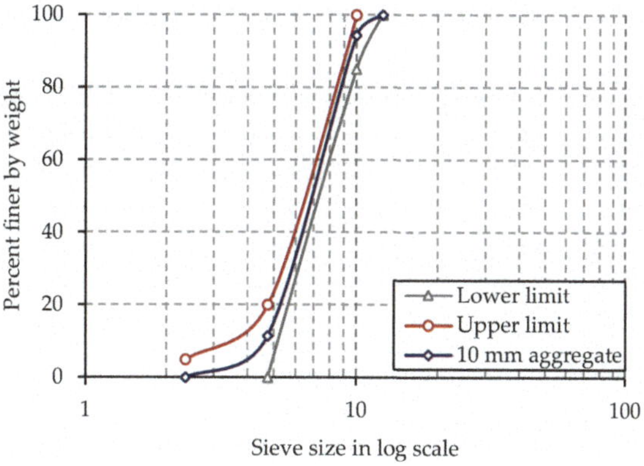

(a) 10 mm coarse aggregate

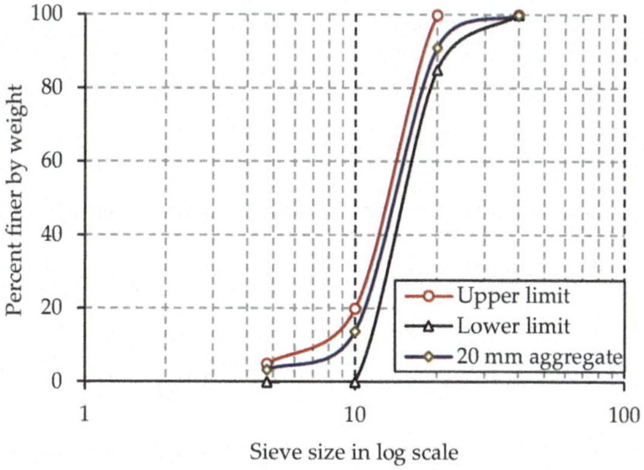

(b) 20 mm coarse aggregate

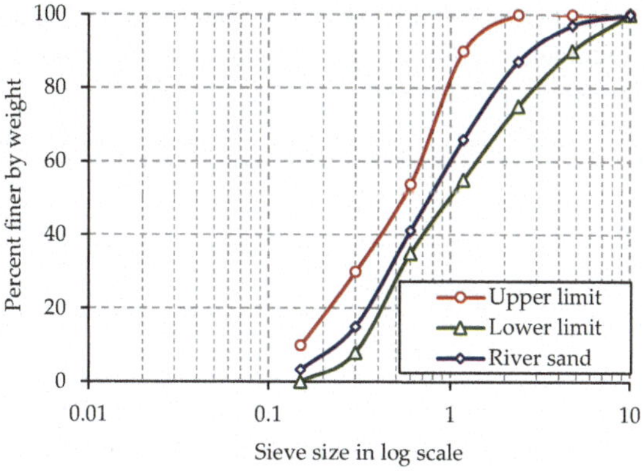

(c) River sand

Table 4 Mix proportioning of cast-in-situ laboratory concrete

Grade	Quantity (kg/m^3)				W/C ratio	Water (kg/m^3)	Superplasticizer (kg/m^3)
	Cement	Fly ash	Sand	CA			
M20	350	–	640	1250	0.50	175	–
M30	380	–	620	1220	0.45	171	–
M40	380	76	600	1190	0.38	173	4.32
M50	420	84	560	1150	0.35	176	5.04
M60	440	88	530	1100	0.34	175	5.28

Table 5 Mix proportioning of pumpable concrete

Grade	Quantity (kg/m^3)						W/C ratio	Water (kg/m^3)	Superplasticizer (kg/m^3)
	Cement	Fly ash	Fine sand <0.25 mm	Sand 0.25 to 2.36 mm	CA				
					10 mm	20 mm			
M25	360	90	150	600	300	650	0.42	184	5.3
M35	375	95	145	580	285	630	0.39	183	5.64
M40	400	100	140	560	277	620	0.36	180	6.0
M45	425	107	136	550	270	610	0.35	186	6.4
M50	440	110	135	540	260	600	0.32	176	6.6

$$\text{Linear form}: Y = A(\text{RN}) + B \tag{2}$$

$$\text{Exponential form}: Y = Ae^{\text{RN}} \tag{3}$$

$$\text{Power form}: Y = A(\text{RN})^{B} \tag{4}$$

where Y is f_c and RN is the rebound hammer number measured during testing. The constants A and B are predicted from the relationship plot between RN and f_c.

Similar to RN, the general forms of correlation between f_c and UPV are also expressed in the following:

$$\text{Linear form}: Y = A(V) + B \tag{5}$$

$$\text{Exponential form}: Y = Ae^{V} \tag{6}$$

$$\text{Power form}: Y = A(\text{V})^{B} \tag{7}$$

where Y is f_c and V is UPV (km/s) measured during testing.

Saha and Amanat [17] predicted the empirical model between f_c and RN using different coarse aggregates and found that the linear methods had a higher correlation coefficient than the exponential and power methods. Most of the research reports selected the linear regression approach for developing the empirical model

Table 6 Model established by various researchers

Reference	Equation	Regression method	Testing method
Shariati et al. [7]	$f_c = 1.7206(\text{RN}) - 26.595$	Linear	SRH
Murthi et al. [2]	$f_c = 16.349e^{0.0267(\text{RN})}$	Exponential	SRH
Shariati et al. [7]	$f_c = 15.533\ V - 34.358$	Linear	UPV
Nepomuceno and Bernardo [6]	$f_c = 0.1921e^{1.2749V}$	Exponential	UPV
Nash't et al. [20]	$f_c = 1.19e^{0.715V}$	Exponential	UPV
Nash't et al. [20]	$f_c = 0.788\ (\text{RN})^{1.03}$	Power	SRH
Kazemi et al. [9]	$f_c = 6.813e^{0.05(\text{RN})}$	Exponential	SRH
Saha and Amanat [17]	$f_c = 2.20(\text{RN}) - 45.67$	Linear	SRH
Saha and Amanat [17]	$f_c = 2.39e^{0.072(\text{RN})}$	Exponential	SRH
Saha and Amanat [17]	$f_c = 0.0033(\text{RN})^{2.57}$	Power	SRH
Odimegwu et al. [18]	$f_c = 0.9229(\text{RN}) - 3.3007$	Linear	SRH
Oke et al. [22]	$f_c = 0.6951(\text{RN}) + 3.7852$	Linear	SRH
Al-Mufti and Fried [21]	$f_c = 0.0096e^{1.734V}$	Exponential	UPV

[6, 7, 17, 18]. Meanwhile, the exponential correlation was also considered for developing similar kind of empirical model [2, 10]. The exponential correlation was mostly followed for predicting f_c from UPV [6, 7, 19]. However, linear approaches were also considered for UPV by researchers [4]. Various empirical models developed by previous research findings are highlighted in Table 6.

3.2 *Relationship Between RN and f_c*

The relationship curves between RN and laboratory specimens are developed and demonstrated in Fig. 6. The linear, exponential, and power models are developed to recognize the best-suited model [1, 12]. The correlation between concrete cube specimens prepared in a laboratory environment and RN was developed with a linear model and found to be f_c = 1.1996 (RN) – 10.466 with a higher correlation coefficient of 0.9934, as shown in Fig. 6a. The same linear relationship was established when pumpable concrete was used and found as f_c = 1.4942 (RN) – 23.941 with the R^2 value of 0.9704. A similar kind of linear relationship was established by Shariati et al. [7] between RN and f_c.

The exponential relationships between RN and laboratory specimens were developed, such as $f_c = 5.8969e^{0.0427(\text{RN})}$ with the R^2 value of 0.9578 for laboratory specimens and $f_c = 10.159e^{0.0318(\text{RN})}$ with the R^2 value of 0.9834 for pumpable concrete, as illustrated in Fig. 6b. Similarly, Fig. 6c depicts the power relationships with a higher correlation coefficient between RN and laboratory specimens and pumpable concrete specimens. Kazemi et al. [9] proposed an exponential correlation, whereas Nash't et al. [20] created an empirical model with power relationship between RN and f_c. The model developed in this investigation was comparable with the existing findings [2, 15, 19, 21].

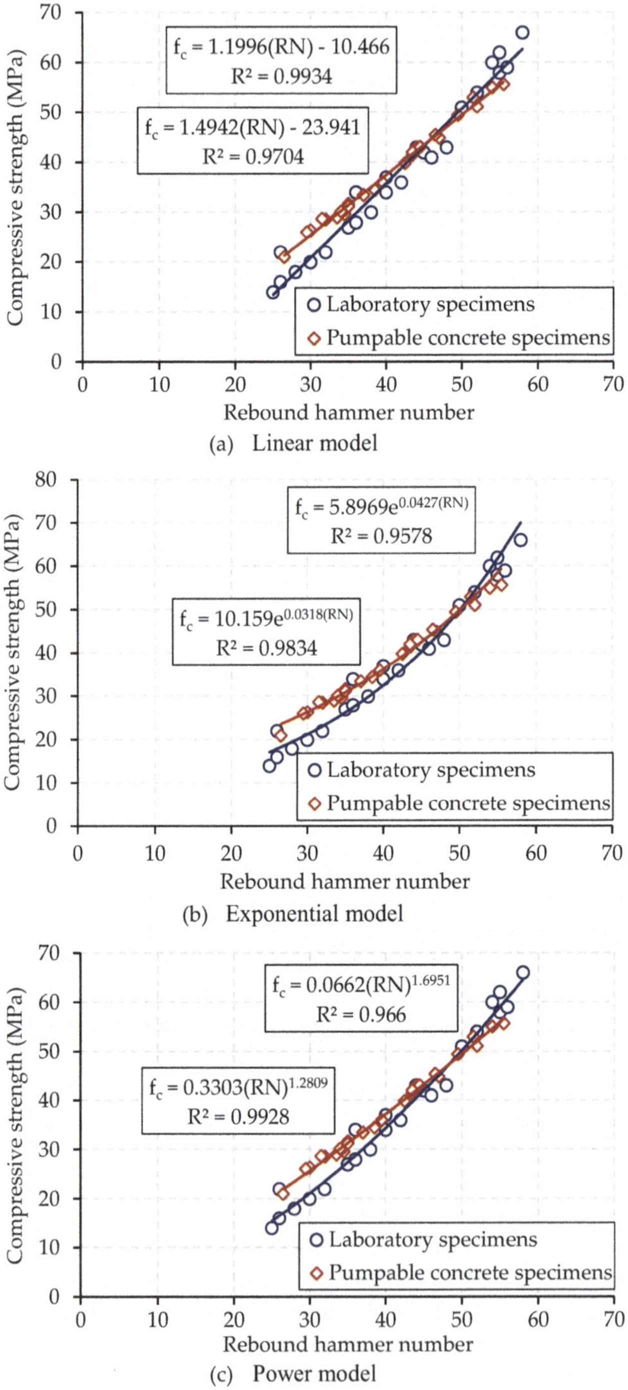

Fig. 6 Relationship between RN and f_c of cast-in-situ laboratory specimens and pumpable concrete specimens: (**a**) linear model, (**b**) exponential model, and (**c**) power model

3.3 Selection of Suitable Model for SRH Test

Poorarbabi et al. [1] predicted an appropriate relationship between destructive and nondestructive results from various models such as power model, linear model, exponential model, and response surface method model. The accuracy of the various models was recognized from the most apt relationship line developed between predicted f_c and actual f_c. The relationship line between predicted f_c using RN tests and actual f_c for the selected three different models is indicated in Fig. 7. The

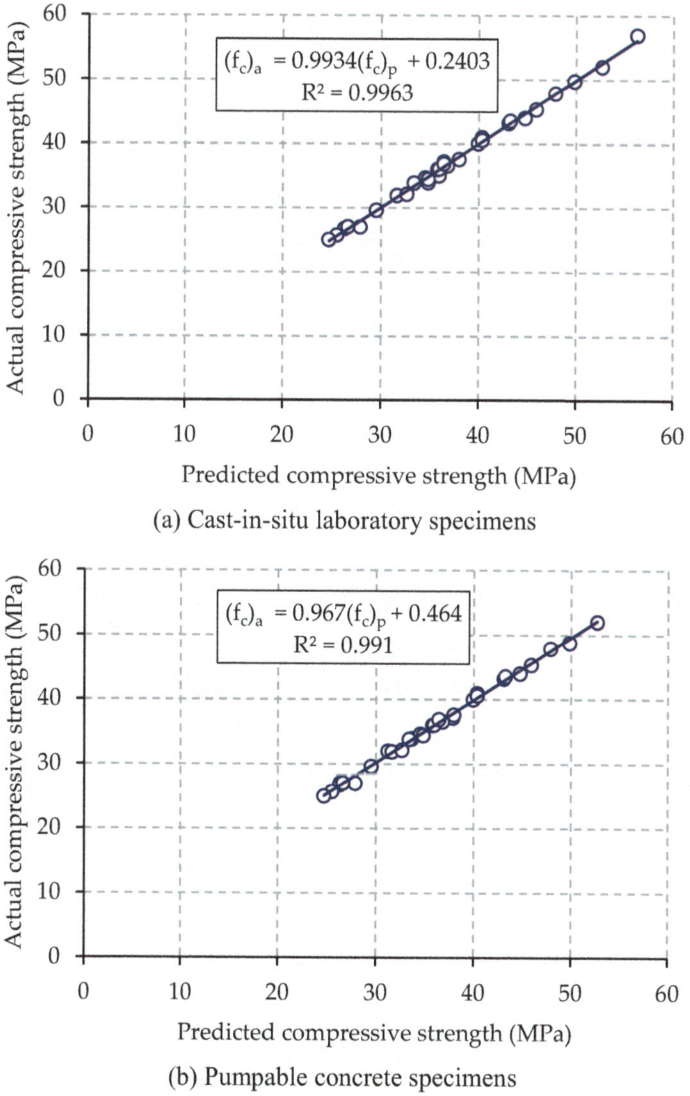

Fig. 7 Relationship between linear model predicted values using RN and tested f_c: (**a**) cast-in-situ laboratory specimens and (**b**) pumpable concrete specimens

deviations between predicted f_c and actual f_c were established from the trend line developed from all the selected models, and the least deviations were found as 0.2403 in cast-in-situ concrete laboratory specimens and 0.464 in pumpable concrete, with higher R^2 values of 0.9963 and 0.991, respectively, for linear model prediction. The deviations of the other two models are displayed in Figs. 8 and 9. The order of accuracy was found to be linear, exponential, and power. Based on this order of accuracy, the linear model was selected when a SRH test was performed to predict f_c of an existing structural concrete.

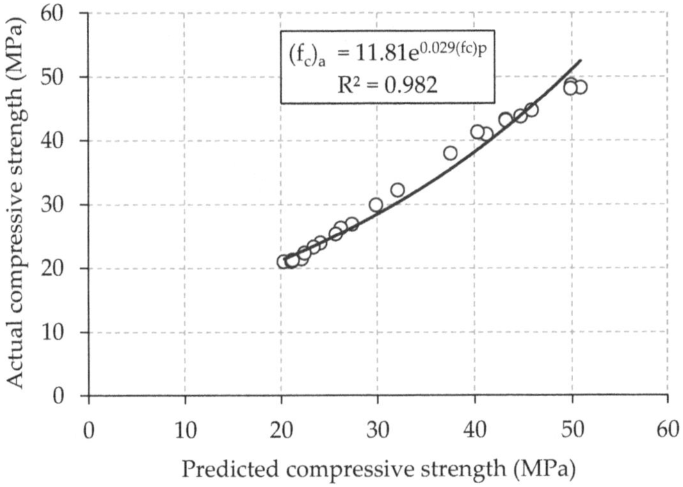

(a) Cast-in-situ laboratory specimens

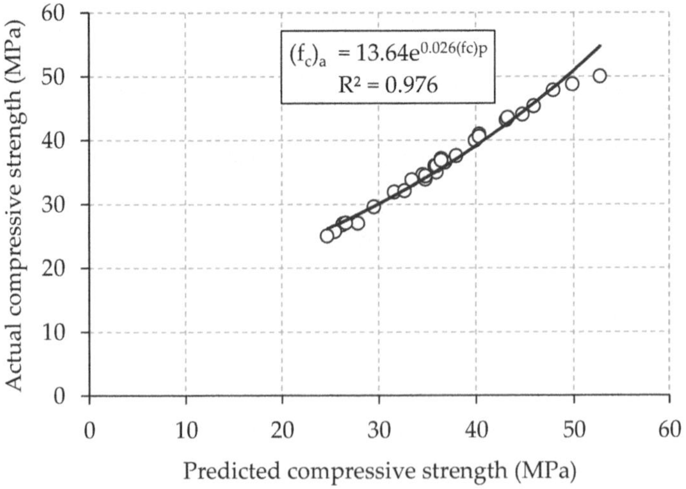

(b) Pumpable concrete specimens

Fig. 8 Relationship between exponential model predicted values using RN and tested f_c: (**a**) cast-in-situ laboratory specimens and (**b**) pumpable concrete specimens

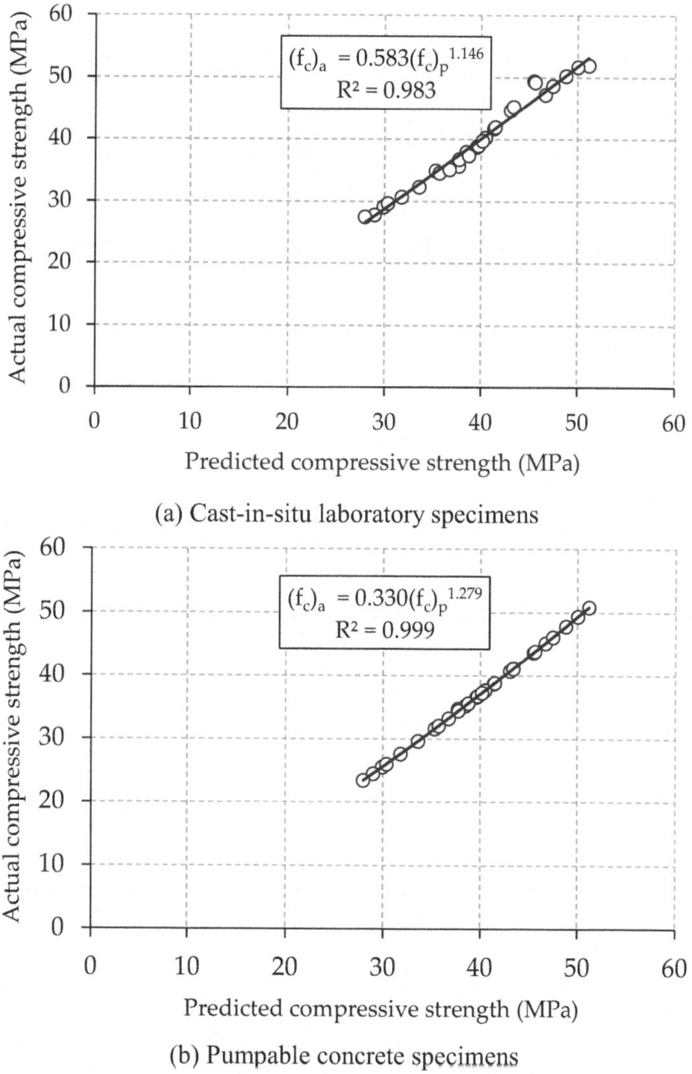

Fig. 9 Relationship between power model predicted values using RN and tested f_c: (**a**) cast-in-situ laboratory specimens and (**b**) pumpable concrete specimens

3.4 Relationship Between UPV and f_c

The correlation curves between UPV and f_c of cast-in-situ laboratory specimens are developed and illustrated in Fig. 10. The best-fit linear model was established with a higher correlation coefficient, as shown in Fig. 10a. The linear model between f_c of cast-in-situ laboratory cube specimens and UPV was developed and found to be

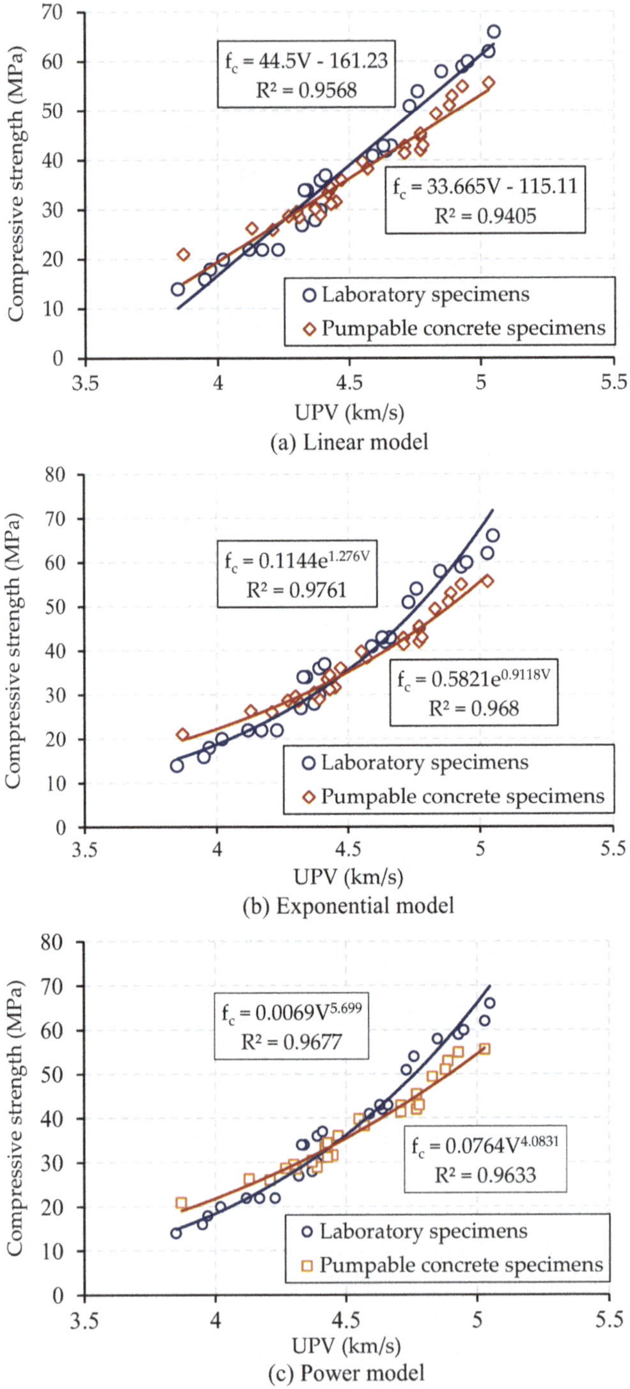

Fig. 10 Relationship between UPV and f_c of cast-in-situ laboratory specimens and pumpable concrete specimens: (**a**) linear model, (**b**) exponential model, and (**c**) power model

$f_c = 44.5\ V - 161.23$ with the R^2 value of 0.9568. A similar linear model between the strength of pumpable concrete cube specimens and UPV was developed, and the equation $f_c = 33.665\ \text{V} - 115.11$ with a correlation coefficient of 0.9405 was found. The exponential relationships between UPV and laboratory specimens were developed, such as $f_c = 0.1144e^{1.276V}$ with the R^2 value of 0.9761 for cast-in-situ laboratory specimens and $f_c = 0.5821e^{0.9118V}$ with the R^2 value of 0.968 for pumpable concrete, as presented in Fig. 10b. A similar exponential relationship was established for normal concrete by Al-Mufti and Fried [21] and Nash't et al. [20]. The power relationships were established with the results obtained in this investigation, as revealed in Fig. 10c, with the R^2 values of 0.9677 and 0.9633 for laboratory specimens and pumpable concrete, respectively. The model developed between f_c and UPV in this investigation was comparable with the previous findings [6, 7, 20, 22].

3.5 Selection of Suitable Model for UPV Test

The accuracy of the various models obtained from UPV test results was recognized by the most apt relationship line developed between predicted f_c and actual f_c [1]. The relationship line between predicted f_c using UPV tests and actual f_c for the selected three different models is demonstrated in Figs. 11, 12, and 13. The deviations between predicted f_c and actual f_c were established from the trend line that developed from all the selected models, and the least deviations were found as 1.2402 in cast-in-situ concrete laboratory specimens and 0.9367 in pumpable concrete specimens, with higher R^2 values of 0.996 and 0.9923, respectively, for linear model prediction, as illustrated in Fig. 11. The correlations of the other two models are indicated in Figs. 12 and 13. The order of accuracy was found to be power, linear, and exponential. Based on this order of accuracy, the power model was selected when a UPV test was performed to predict f_c of an existing structural concrete.

3.6 Assessing On-Site fc of Concrete Structures

The next stage included assessing the on-site compressive strength using nondestructive methods (shown in Figs. 2 and 4), with the aid of models developed from this investigation. The 28-day f_c of the cubes prepared during the pouring of concrete in on-site structural elements was tested and preserved for comparison purposes. The linear model obtained from the SRH test results and the power model achieved from the UPV test results were employed to calculate f_c of concrete in the existing structure, as depicted in Table 7. The investigation results sourced from four different structures are presented in Table 8 in order to compare the predicted strength of on-site elements with the actual f_c of both normal and high-strength and

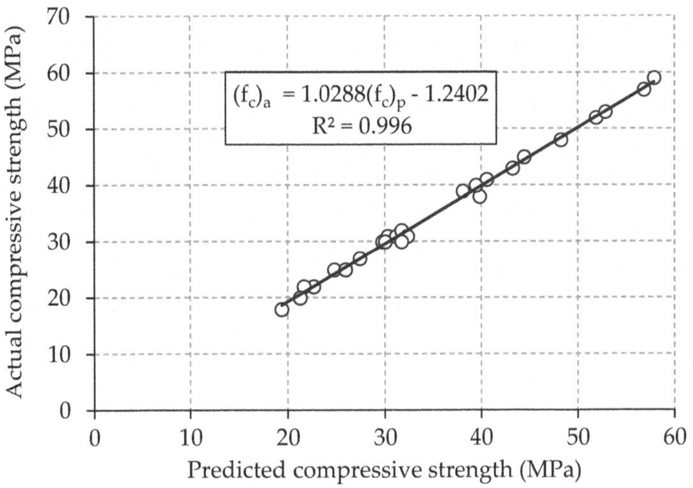

(a) Cast-in-situ laboratory specimens

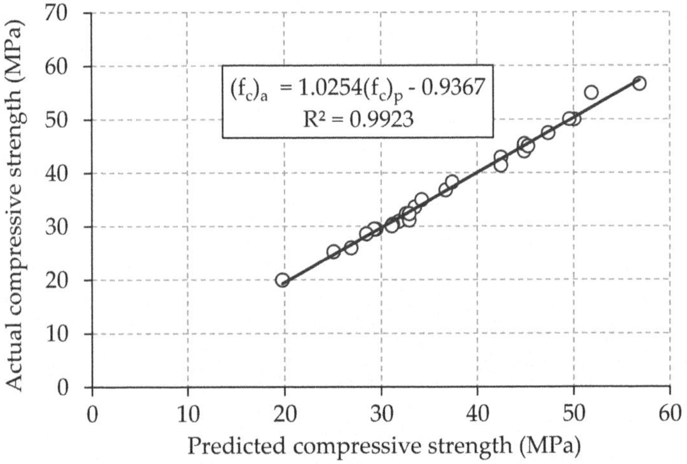

(b) Pumpable concrete specimens

Fig. 11 Relationship between linear model predicted values using UPV and tested f_c: (**a**) cast-in-situ laboratory specimens and (**b**) pumpable concrete specimens

pumpable concretes. It can be noted that the predicted strength values are closer to the actual strength obtained from the laboratory investigation.

The model predicted from the investigation based on selected existing on-going structures can be preserved as an additional document of the structures and can be considered for performance evaluation during the future service period. It is also suitable for the health monitoring of the structures.

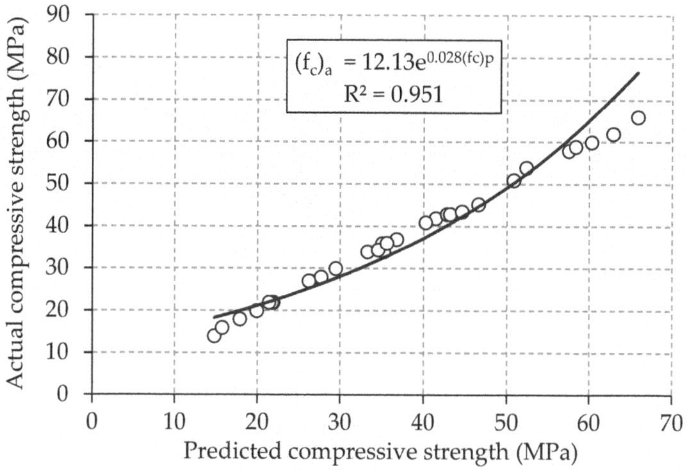

(a) Cast-in-situ laboratory specimens

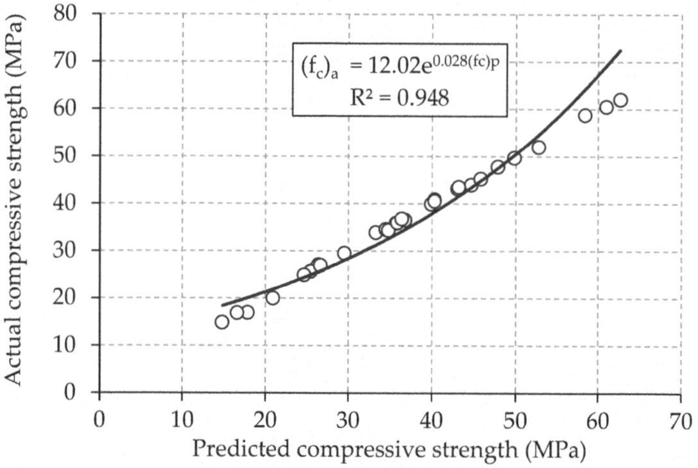

(b) Pumpable concrete specimens

Fig. 12 Relationship between exponential model predicted values using UPV and tested f_c: (**a**) cast-in-situ laboratory specimens and (**b**) pumpable concrete specimens

4 Conclusions

Two different categories of concrete, namely, (1) cast-in-situ concrete and (2) pumpable concrete, were selected to establish the empirical models to assess f_c using NDTs. The following conclusions were drawn with reference to the three different forms of models in each test:

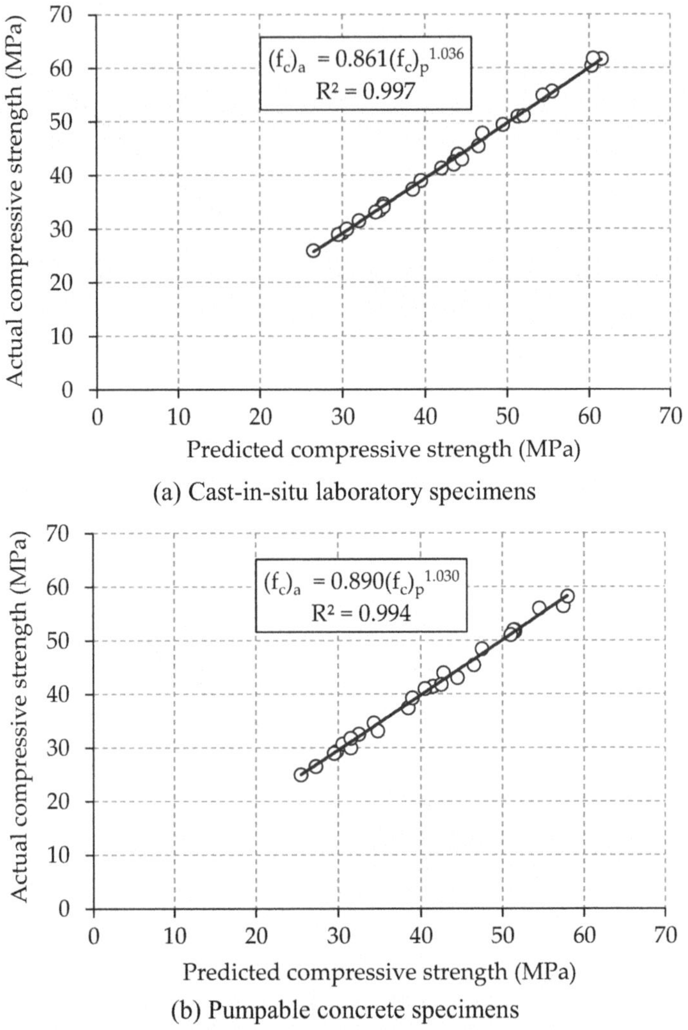

Fig. 13 Relationship between power model predicted values using UPV and tested f_c: (**a**) cast-in-situ laboratory specimens and (**b**) pumpable concrete specimens

Table 7 Selected empirical models from this investigation

Type of test	Type of concrete	Type of model	Empirical model	R^2 value
SRH	Cast-in-situ concrete	Linear model	$f_c = 1.1996$ (RN) – 10.466	0.9934
	Pumpable concrete		$f_c = 1.4942$ (RN) – 23.941	0.9704
UPV	Cast-in-situ concrete	Power model	$f_c = 0.0069V^{5.699}$	0.9677
	Pumpable concrete		$f_c = 0.0764V^{4.0831}$	0.9633

Table 8 Comparison between actual and predicted f_c of on-site tests

Testing method	On-site location	Cast-in-situ concrete			Pumpable concrete		
		Actual concrete cube strength	Predicted strength	Difference (%)	Actual concrete cube strength	Predicted strength	Difference (%)
SRH test	1	38.37	37.41	−2.50	28.59	29.37	2.73
	2	41.48	42.05	1.37	53.04	52.77	−0.51
	3	29.04	30.17	3.89	42.96	43.17	0.49
	4	33.48	32.63	−2.54	36.07	36.55	0.13
UPV test	1	51.57	52.69	2.17	44.51	43.79	−1.62
	2	49.57	49.98	0.83	39.53	40.13	1.52
	3	29.15	28.82	−1.12	42.57	43.73	2.72
	4	33.53	32.73	−2.39	54.13	53.83	−0.55

- Linear model was suggested to assess the performance of concrete structure while using RN as the variable parameter in the empirical model and developed as f_c = 1.1996 (RN) − 10.466 for cast-in-situ concrete and f_c = 1.4942(RN) - 23.941 for pumpable concrete with higher coefficients of correlation of 0.9934 and 0.9704, respectively.
- Power model was suggested to assess the performance of concrete structure when UPV was considered, and the developed models are $f_c = 0.0069V^{5.699}$ for cast-in-situ concrete and $f_c = 0.0764V^{4.0831}$ for pumpable concrete with higher coefficients of correlation of 0.9677 and 0.9633, respectively.
- It is evident that the exponential models found in the references are not adoptable to the present findings used in pumpable concrete in the selected region.
- The appropriate strength performance of the existing structure can be predicted using the selected empirical models.
- These equations are recommended to construction engineers and clients for assessing the structural performance in the future.

References

1. A. Poorarbabi, M. Ghasemi, M.A. Moghaddam, Concrete compressive strength prediction using non-destructive tests through response surface methodology. Ain Shams Eng. J. **11**, 939–949 (2020). https://doi.org/10.1016/j.asej.2020.02.009
2. P. Murthi, K. Poongodi, R. Gobinath, Correlation between rebound hammer number and mechanical properties of steel fibre reinforced pavement quality concrete. Mater. Today Proc. **39**(7), 142 (2020). https://doi.org/10.1016/j.matpr.2020.06.402
3. M. Ju, K. Park, H. Oh, Estimation of compressive strength of high strength concrete using non-destructive technique and concrete core strength. Appl. Sci. **7**(12), 1249 (2017). https://doi.org/10.3390/app7121249
4. D. Breysse, Nondestructive evaluation of concrete strength: an historical review and a new perspective by combining NDT methods. Constr. Build. Mater. **33**, 139–163 (2012). https://doi.org/10.1016/j.conbuildmat.2011.12.103

5. IS: 13311–1992 (Part-1), *Non-Destructive Testing of Concrete—Methods of Test. Part 1. Ultrasonic Pulse Velocity* (Bureau of Indian Standards, New Delhi)
6. M.C.S. Nepomuceno, L.F.A. Bernardo, Evaluation of self-compacting concrete strength with non-destructive tests for concrete structures. Appl. Sci. **9**, 5109 (2019). https://doi.org/10.3390/app9235109
7. M. Shariati, N.H. Ramli-Sulong, M.M. Arabnejad, P. Shafigh, H. Siaei, Assessing the strength of reinforced concrete structures through ultrasonic pulse velocity and Schmidt rebound hammer tests. Sci. Res. Essays **6**(1), 213–220 (2011). https://doi.org/10.5897/SRE10.879
8. W. Martinez-Molina, A.A. Torres-Acosta, J.C. Jauregui, H.L. Chavez-Garcia, E.M. Alonso-Guzmam, M. Graff, J. Arteaga-Arcos, Predicting concrete compressive strength and modulus of rupture using different NDT techniques. Adv. Mater. Sci. Eng. **2014**, 742129 (2014). https://doi.org/10.1155/2014/742129
9. M. Kazemi, R. Madandoust, J.D. Brito, Compressive strength assessment of recycled aggregate concrete using Schmidt rebound hammer and core testing. Constr. Build. Mater. **224**, 630–638 (2019). https://doi.org/10.1016/j.conbuildmat.2019.07.110
10. K. Ali-Benyahia, Z. Sbartaï, D. Breysse, M. Ghrici, S. Kenai, Improvement of non-destructive assessment of on-site concrete strength: Influence of the selection process of cores location on the assessment quality for single and combined NDT techniques. Constr. Build. Mater. **195**, 613–622 (2019). https://doi.org/10.1016/j.conbuildmat.2018.10.032
11. Z.M. Sbartaï, M. Alwash, D. Breysse, Combining the bi-objective approach and conditional coring for a reliable estimation of on-site concrete strength variability. Mater. Struct. **54**, 230 (2021). https://doi.org/10.1617/s11527-021-01820-9
12. D. Breysse, X. Romao, M. Alwash, Z.M. Sbartaï, V.A.M. Luprano, Risk evaluation on concrete strength assessment with NDT technique and conditional coring approach. J. Build. Eng. **32**, 101541 (2020). https://doi.org/10.1016/j.jobe.2020.101541
13. IS: 13311–1992 (Part-2), *Non-Destructive Testing of Concrete—Methods of Test. Part 1. Rebound Hammer Test* (Bureau of Indian Standards, New Delhi)
14. IS: 269–2015, *Specification for Ordinary Portland Cement, 6th Revision* (Bureau of Indian Standards, New Delhi)
15. IS: 383–1970, *Coarse and Fine Aggregate for Concrete—Specification* (Bureau of Indian Standards, New Delhi)
16. ASTM C-494, *Standard Specification for Chemical Admixture for Concrete* (ASTM International, Washington, DC)
17. A.S. Saha, K.M. Amanat, Rebound hammer test to predict in-situ strength of concrete using recycled concrete aggregates, brick chips and stone chips. Constr. Build. Mater. **268**, 121088 (2021). https://doi.org/10.1016/j.conbuildmat.2020.121088
18. T.C. Odimegwu, A.B.M.A. Kaish, I. Zakaria, M.M. Abood, M. Jamil, K.O. Ngozi, Nondestructive determination of strength of concrete incorporating industrial wastes as partial replacement for fine aggregate. Sensors **21**(24), 8256 (2021). https://doi.org/10.3390/s21248256
19. O.D. Atoyebi, O.P. Ayanrinde, J. Oluwafemi, Reliability comparison of Schmidt rebound hammer as a non-destructive test with compressive strength tests for different concrete mix. J. Phys. Conf. Ser. **1378**, 032096 (2019). https://doi.org/10.1088/1742-6596/1378/3/032096
20. I.H. Nash't, S.H. A'Bour, A.A. Sadoon, Finding a unified relationship between crushing strength of concrete and non-destructive tests, in *3rd MENDT—Middle East Nondestructive Testing Conference & Exhibition*, (2005)
21. R.L. Al-Mufti, A.N. Fried, The early age non-destructive testing of concrete made with recycled concrete aggregate. Constr. Build. Mater. **37**, 379–386 (2012). https://doi.org/10.1016/j.conbuildmat.2012.07.058
22. D.A. Oke, G.F. Oladiran, S.B. Raheem, Correlation between destructive compressive testing (DT) and non destructive testing (NDT) for concrete strength. Int. J. Eng. Res. Gen. Sci **3**(5), 27–30 (2017)

Engineered Cementitious Composites for Sustainable and Long-Lasting Infrastructures

Muhammad Hammad, Alireza Bahrami, Sikandar Ali Khokhar, Ameer Hamza, and Fazal Rehman

1 Introduction

Engineered cementitious composites (ECCs) are a type of cementitious material from the family of fiber-reinforced concrete (FRC), renowned by their extraordinary ductile behavior and strain-hardening behavior in uniaxial tension [1, 2]. ECCs demonstrated a tensile strain capacity (1–3%) along with multiple microcracking, indicating their superiority in mechanical and durability properties [3]. These unique attributes enable ECCs to undergo significant deformation (due to several fine cracks) and additional load-bearing capacity even after the initiation of cracks, which is absent in conventional concrete and many FRC variants [4].

ECCs represent a family of cementitious composites that prioritizes ductility and higher tensile strength. The design philosophy of ECCs is enormously different from other special kinds of concrete like high-strength concrete and ultra-high

M. Hammad
School of Civil Engineering, University of Nottingham, University Park, Nottingham, UK

A. Bahrami (✉)
Department of Building Engineering, Energy Systems and Sustainability Science, Faculty of Engineering and Sustainable Development, University of Gävle, Gävle, Sweden
e-mail: alireza.bahrami@hig.se

S. A. Khokhar
Ben Lyles School of Civil Engineering, Purdue University, West Lafayette, IN, USA

Bendcrete Construction Services (Pvt) Ltd, National Science and Technology Park, Sector H-12, Islamabad, Pakistan

A. Hamza · F. Rehman
Bendcrete Construction Services (Pvt) Ltd, National Science and Technology Park, Sector H-12, Islamabad, Pakistan

School of Civil and Environmental Engineering, NUST Institute of Civil Engineering, National University of Sciences and Technology, Sector H-12, Islamabad, Pakistan

A. Bahrami (ed.), *Advances in Sustainable Concrete for Construction*,
https://doi.org/10.1007/978-3-031-85052-3_5

performance concrete, which primarily focus on denser packing density [4, 5]. Alternatively, the design of ECCs is based on their micromechanical model, which enforces that the microstructure be precisely engineered to foster synergistic interactions among their constituents, including fibers, matrix, and fiber/matrix interfaces [6]. This method, guided by micromechanics of ECCs, stands out better, highlighting their custom design rules to change the strain-softening behavior into strain-hardening. The constitutive model of ECCs is similar to that of metallic materials, having a different yield strength and then a strain-hardening behavior with lots of load drops displaying the formation of several cracks [7]. This strain-hardening behavior of ECCs shows how adaptable they are for structures where ductility and robustness are super important. ECCs' tensile ductility is at least two times higher than that of traditional concrete, while their compressive strength (f'_c) can be more than 200 MPa, depending on specific design requirements [1, 8]. The typical constitutive model of ECC with the maximum breadth of the crack is illustrated in Fig. 1.

The microcracking resistance properties of ECCs contribute to their suitability for enhancing the resilience, durability, and sustainability of the structures. By reducing the chance of sudden fracture failures, ECCs help increase the service life and boost the performance of infrastructure. Research has proven that new limestone calcined clay cement (LC3)-based low-carbon ECC showed some impressive improvements, especially in terms of f'_c, self-healing effects, and microcracking, with almost 62% less carbon footprints than traditional ECC [10]. Focusing on ductility and crack width control really highlights how ECCs could change the way we build, paving the way for stronger and more sustainable infrastructure.

In this chapter, we take a deep look into how ECCs fit into sustainable infrastructures. The discussion covers topics like sustainable building materials [11–13], structures, and systems (SIMSS) approach [9, 14]. This study presents an analysis of methodologies related to life cycle assessment (LCA) of structures' sustainability, alongside strategies for improving the environmental performance of ECCs. Furthermore, the sustainable commercial viability of ECCs in the context of buildings, transport networks, and energy-absorbing systems is thoroughly studied.

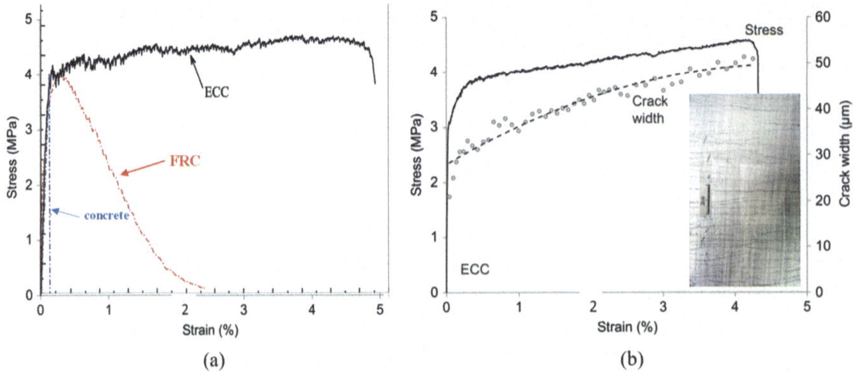

Fig. 1 Typical tensile behavior of ECCs: (a) stress–strain curve and (b) crack width [9]

Based on these findings, detailed recommendations for further study are discussed to help researchers and engineers in the effective use of ECCs for the advancement of resilient buildings.

2 Sustainability of ECC Infrastructures

Civil infrastructures, such as buildings, roads, and bridges, make up an important part of the modern world and notably impact the quality of life globally. However, the conventional methods used for designing and constructing such structures often ignore essential sustainability parameters. This oversight then leads to adverse impacts on economy and global environment throughout their life [15, 16].

The scale and service life requirements of civil infrastructures necessitate a shift toward sustainable design and management practices to address the challenges of resource depletion, environmental degradation, and societal resilience [17–19]. As urban populations continue to increase quickly, the requirement for buildings, particularly in highly populated urban centers, increases, exacerbating material consumption and environmental impact.

Concrete, the most widely utilized construction material, accounts for a substantial portion of global material consumption, with approximately 12 billion tons consumed annually [20]. However, manufacturing of cement, the key ingredient in concrete, produces large amounts of carbon dioxide, which in turn has a negative impact on climate change and pollution [21, 22].

While the material production phase represents a significant portion of infrastructure-related carbon emissions, the usage phase, characterized by maintenance and repair activities, often dominates the life cycle impact of infrastructures [23]. Less maintenance, when combined with low-quality materials and bad design practices, results in frequent repairs and replacements of structures, leading to further utilization of resources and energy, and increases pollution [24].

The deteriorating condition of infrastructure systems poses multifaceted challenges, including water loss due to leakage, energy inefficiency in building operations, and increased vulnerability to extreme environmental hazards. To tackle these problems, a proper approach is necessary to consider sustainability principles throughout the life cycle of the infrastructure, from material selection and design to construction, operation, and maintenance [17, 25].

ECCs offer a viable solution to many of the sustainability-related issues for civil structures. With their extraordinary ductility, crack control, and durability, ECCs can increase the life cycle of structures, reduce the need for excess repair works, and tackle environmental issues [12, 13, 26]. By being resilient, having high performance, and using less resources, ECCs are contributing toward the realization of a more robust and eco-friendlier built environment for present and future generations.

2.1 SIMSS Design Approach

A special approach to improve a structure's sustainability is the SIMSS design approach, as suggested by Lepech [14]. This framework includes materials optimization, structure design, and infrastructure system functions with determination of life cycle, as displayed in Fig. 2.

The SIMSS design approach focuses on the relationship of the different steps of a structure's life cycle and their participation toward sustainability factors. It indicates those decisions made at the materials scale, such as the mix design and microstructure of composites like ECCs, impacting performance and contributing to system-level sustainability throughout the service life of structure.

At the system level, maintenance of structural serviceability is important, as it requires consideration for resource inputs and emissions outputs [27]. The durability of structures under mechanical and environmental influence decides the repairs and maintenance requirement at the structural level [28]. These three levels are somehow connected, with materials and structural properties impacting each other across different levels.

Moreover, the SIMSS approach considers a linked timescale, where intentional microstructure adjustment leads to ECCs that allow quicker construction process, improve durability, and enhance service life. This perspective highlights how crucial it is to think about long-term sustainability when making choices about materials and structural design [29, 30].

The evaluation module within the SIMSS framework encompasses tools such as LCA and economic modeling to assess infrastructure sustainability across various phases of the life cycle. This includes material development, construction, operation, restoration, and lifespan considerations. The result of this evaluation helps

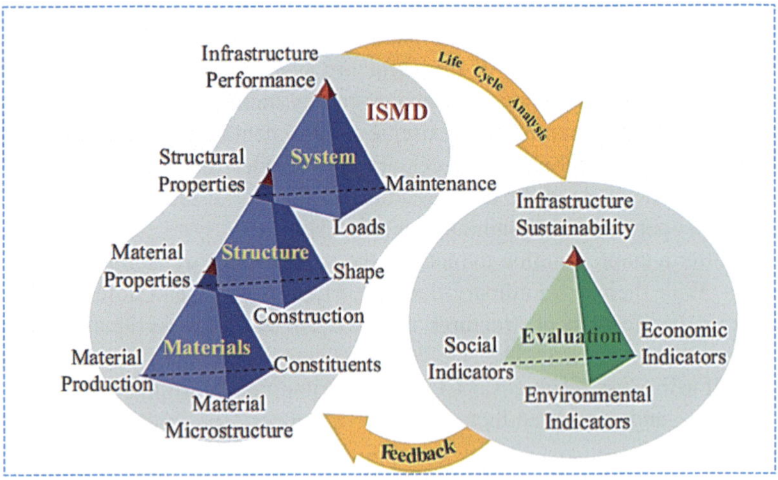

Fig. 2 Sustainable infrastructure materials, structures, and systems design approach [14]

iterative designs of infrastructure systems at all levels, optimizing performance and cost-effectiveness [31, 32].

However, the current state of infrastructure design often falls short of the SIMSS approach. Structural design tends to prioritize f'_c of concrete, ignoring important material properties such as tensile cracking patterns. This disconnection between materials, structure, and system development limits the chances for system optimization and affects the sustainability factors [13]. Furthermore, evaluations generally focus only on economic factors, ignoring social and environmental considerations, thus making it harder to achieve desired sustainability standards.

2.2 *Infrastructure Sustainability LCA*

LCA is necessary in determining the sustainability of ECCs. It aids in pointing out areas where amendments can be made by examining the entire lifespan of materials. Not only does LCA pinpoint areas for improvements, but it also reveals the negative influence of a few materials, helping us avoid adverse effects on our environment [19, 33]. Another objective of LCA is to help design (optimize) ECCs for better environmental performance [34]. Figure 3 illustrates how LCA can play a role in the optimal design of ECCs. Sustainability metrics can hint at changes in material development, which might affect ECCs' features and possibly improve how they perform structurally.

One of the important goals of conducting LCA is to aid policymakers in making important decisions about utilizing novel materials. Novel construction materials

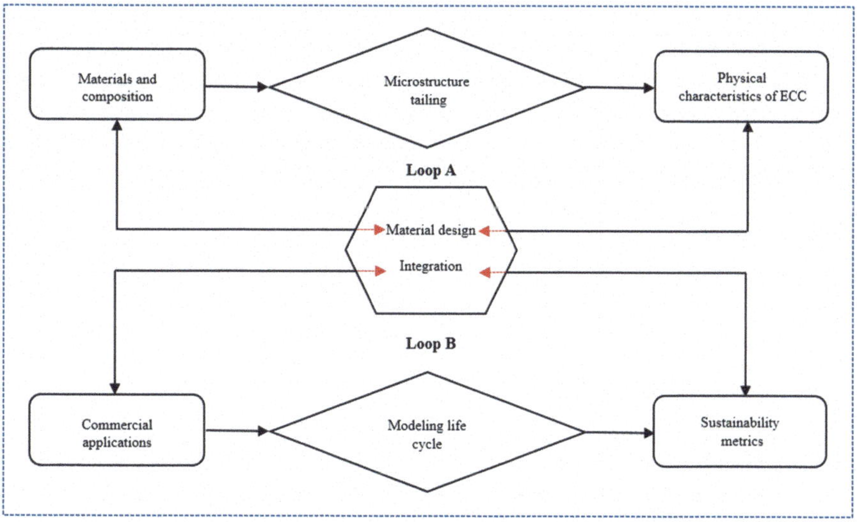

Fig. 3 LCA loops for optimal design of materials [5]

often face difficulties like a lack of guidelines and higher initial costs [34]. Lepech's model, for instance, suggests that reinforced ECC members with 3% tensile strain capacity will have a service life of about 60 years, while ECC members with 2% strain capacity will have a lifespan of fewer than 35 years [14]. This demonstrates the compromise between the price of initial materials and ongoing maintenance costs. LCA offers a wider point of view, exhibiting the long-term socioeconomic and environmental benefits that might not be clear when only considering initial material costs. By considering these metrics, LCA helps balance the potential benefits against the risks of using novel materials [35]. It is important to note that the outcomes of a comparative LCA depend not only on the constituents being used but also on the site, as well as on the type of structural system, including traffic patterns and climate conditions.

Hou et al. [36] conducted experiments and LCA of green and durable ECC (GD-ECC) focused on potential application in bridge deck construction, which utilized low-carbon raw materials, including reused polyethylene fiber, along with a carbonation curing process. The outcomes revealed that GD-ECC had an impressive tensile strength of 3.5 MPa and tensile ductility of 7.7%. Despite its eco-friendly mixed design, the carbon footprints and price of GD-ECC were similar to those of traditional concrete; this could be attributed to the use of recycled fibers and desert sand. Also, reinforced GD-ECC displayed more resistance to corrosive environment, having a life that is about two times as long as reinforced mortar. In addition, LCA of GD-ECC showed a big drop in both costs and carbon footprints when it is compared with conventional ECC, displaying its promise for eco-friendly infrastructure development. Moreover, figuring out the degradation rate in LCA is very important for knowing how long structures will stand, especially for those that need many repairs over time. Novel composites like ECCs can affect the durability of structures and how frequently they need to be repaired, although statistical information on ECC degradation is often lacking. As a result, modeling of service life becomes important. Steel rebar corrosion is a typical source of deterioration in reinforced cement concrete (RCC), especially in cold and humid areas. Although concrete often acts as a barrier, exposure to salt can cause chloride ions to enter and start corrosion, resulting in the formation of rust and ultimately cracks in concrete. While concrete cracking frequently reduces the efficiency of corrosion control techniques, their goal is to postpone the onset of corrosion. Because of its limited crack width and steady deformation state, ECCs may provide more accurate corrosion propagation time estimates than traditional concrete, as depicted in Figs. 4 and 5 [37, 38].

2.3 Greening of ECCs

As the building industry is becoming increasingly concerned about sustainability, the pace of research on the greening of ECCs has also increased [39]. ECCs possess the capacity to decrease the power required for the constructed infrastructure because of being self-healing and crack resistant. Multiple attempts have been made

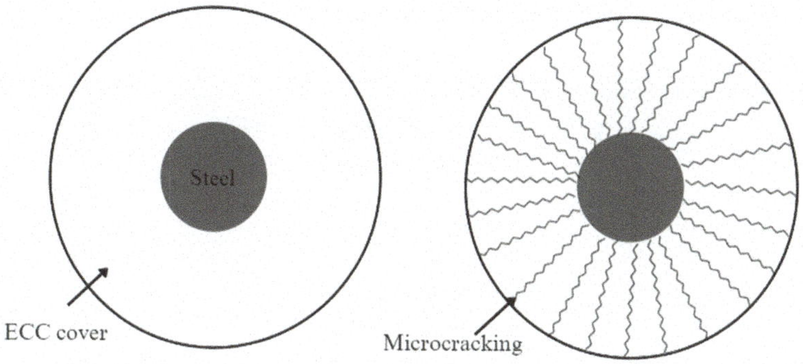

Fig. 4 Expansion of actively corroding steel accommodated by inelastic straining of ECCs [5]

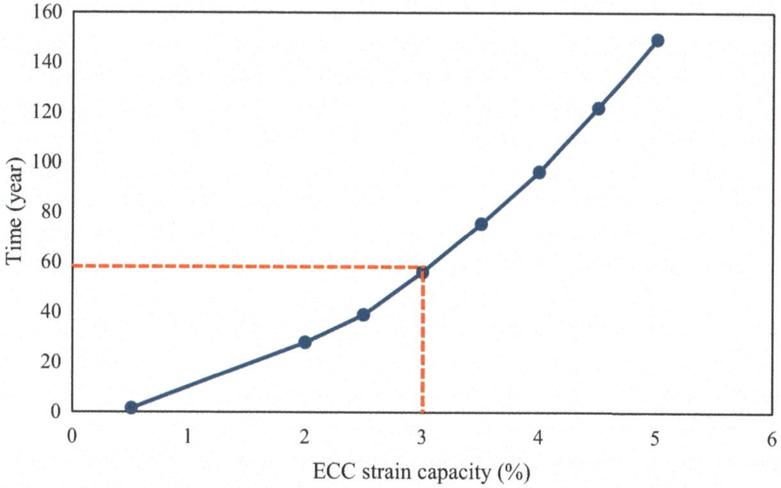

Fig. 5 Estimated service life of ECC members after de-passivation of rebars [5]

to produce greener ECCs in the last few years [40–45]. The greener fibers, fillers, and binders have been employed by various researchers to achieve the desired outcomes, as listed in Table 1. For instance, as a replacement for cement, supplementary cementitious materials (SCMs) such as fly ash (FA), ground granulated blast furnace slag (GGBFS), silica fume (SF), kiln dust, and limestone calcined clay (or metakaolin) have a lower environmental impact and therefore improve the sustainability of cementitious binders in concrete [40–42, 46, 47]. Specifically, FA is employed extensively in ECCs and provides notable embodied carbon reductions. Yu et al. [42, 47] suggested that replacing cement with 80% FA can result in an approximate 70% decrease in total hydration heat, a 15% decrement in material production costs, a 70% decrease in CO_2 emissions, and a 60% decrement in embodied energy for a cement-based concrete that aims for f'_c of 45 MPa. Wu et al. [48] reported that the environmental impacts of ECCs with complete cement substitution

Table 1 Various low-carbon binders and fillers used in ECCs

Binders (fillers)	Low-carbon binders (fillers)	References
Binders (non-cement-based)	MgO and FA	[50, 51]
	Calcium sulfoaluminate and FA	[52–56]
	Limestone calcined clay and FA	[57–59]
Cement-based	FA	[42, 47, 60, 61]
	GGBFS and FA	[42, 60–62]
	SF and FA	[63–65]
	Wasted ceramic and FA	[66, 67]
	Rice husk ash and FA	[45, 68]
Fillers	IOT and coarser particles	[40, 69]
	Rice husk ash	[45, 70, 71]
	Hollow glass microspheres	[46, 72, 73]

by half magnesium oxide (MgO) and half FA were superior to those with cement replaced by 70% MgO and 30% FA. Even though MgO has a greater embodied CO_2 (1.4 kg CO_2/kg) than cement (0.85 kg CO_2/kg), CO_2 sequestration through carbonation curing allows for a reduction in the net CO_2 of MgO-based ECCs when compared to cement-based ECCs. Huang et al. [40] investigated the possibility of producing greener ECCs by utilizing iron ore tailings (IOTs). ECCs demonstrate enhanced environmental sustainability without sacrificing mechanical characteristics by substituting some of the cement with IOTs. Comparing the findings to conventional ECCs, a notable decrement in CO_2 (29–63%) and energy consumption (10–32%) was observed. Furthermore, Hou et al. [36] examined the low-cost, low-carbon substitute for ordinary concrete in the form of a GD-ECC. With the use of carbonation curing, desert sand, and recycled polyethylene fiber, GD-ECC presented excellent corrosion resistance, minimum cracking, and good tensile qualities, and its LCA exhibited a 50% reduction in material cost and carbon footprint for a bridge deck application. In another research, they assessed the use of wasted crumb rubber and SF to improve the sustainability of polypropylene fiber-reinforced ECCs [49]. Crumb rubber decreased the microcrack width and enhanced strain hardening; however, it adversely affected f'_c. Although they widened the cracks, SF and a lower water-to-binder ratio improved f'_c. Additionally, ECC's operational and embodied carbon footprint decreased when crumb rubber and a low-carbon binder (LC3) were used together.

3 Commercial Applications of ECC Infrastructures

Although ECCs are uneconomical and have an adverse influence on the environment due to the high fiber costs, they exhibit excellent mechanical properties. After the initial cracking, ECCs show hardening behavior and reach a tensile strain capacity of about 5%, which is nearly 100-folds more than typical concrete. In addition, they have a high modulus of rupture, a high tensile strength of approximately 10 MPa, and f'_c of up to 80 MPa. ECCs also illustrate superior post-crack strength and ductility due to the fiber crack bridging phenomenon. Besides these properties, ECCs are a lightweight, durable, and self-healing composite. Due to these phenomenal properties, ECCs are being utilized worldwide in many structural applications such as retrofit material, coupling beams, shotcrete, pavement, link slabs, etc. A few present commercial applications of ECCs include the production of precast members in plants and on-site applications through normal and self-consolidated casting, as well as spraying (shotcrete, etc.) techniques. Some of the applications of ECCs are discussed in the following section.

3.1 Building Structures

3.1.1 ECC Coupling Beams

Significant shear reinforcement is typically required to assure the safety and flexibility of RCC structures under loading conditions [74, 75]. This reinforcement aids in improving shear resistance, preventing splitting failure and buckling of tensile reinforcement. However, ECCs have been effectively applied in crucial shear zones of structural elements, for example, beam–column joints and infill panels in frame structures, because of their outstanding tension and shear capabilities [76]. Research reveals that transverse reinforcement required in beams can be decreased or even neglected when ECCs are used in place of conventional concrete [10]. Because reinforced ECC beams no longer require adherence to minimum transverse (shear) reinforcement ratios, this frees up more design flexibility in engineering. Furthermore, ECC coupling beams, which are integrated into the core walls of high-rise buildings, offer a promising solution by effectively distributing heavy loads and absorbing seismic energy. The exceptional damage tolerance and ductility of ECCs, in contrast to traditional RCC coupling beams, reduce building repair costs and time [77]. Experiments performed by Kanda et al. [78] demonstrated the capacity of ECCs to sustain seismic forces without experiencing significant strength reduction or shear failures, highlighting the material's resilience under cyclic shear loading. Additionally, ECC coupling beams were utilized in high-rise buildings like Glorio Tower, Nabule Yokohama Tower, and Kitahama Tower. Interestingly, these buildings did not suffer any damage from the seismic shaking caused by far-off earthquakes, proving the effectiveness of ECC coupling beams in seismic-resistant building designs [79]. Moreover, Zhang et al. [80] investigated eight large-scale ECC and RCC coupling beam samples for cyclic loading, as depicted in Fig. 6. Several parameters were considered, such as energy dissipation, stiffness and strength deterioration, and failure patterns. Based on the analysis of these specimens, an equation for shear strength was developed. According to the outcomes, ECC coupling beams with different reinforcements performed better. It also

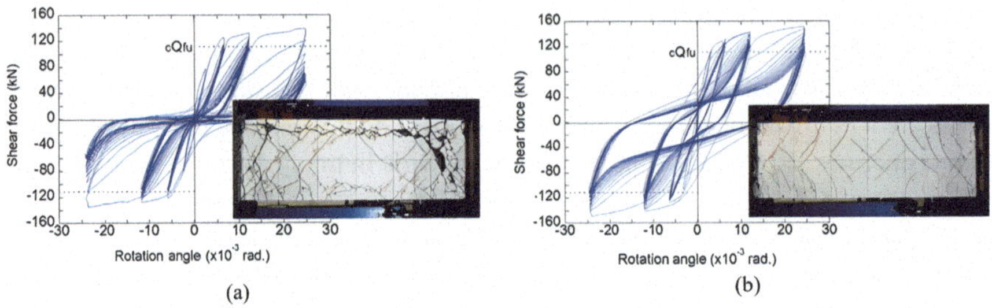

Fig. 6 Hysteretic behavior of coupling beams: (**a**) RCC and (**b**) ECC [5]

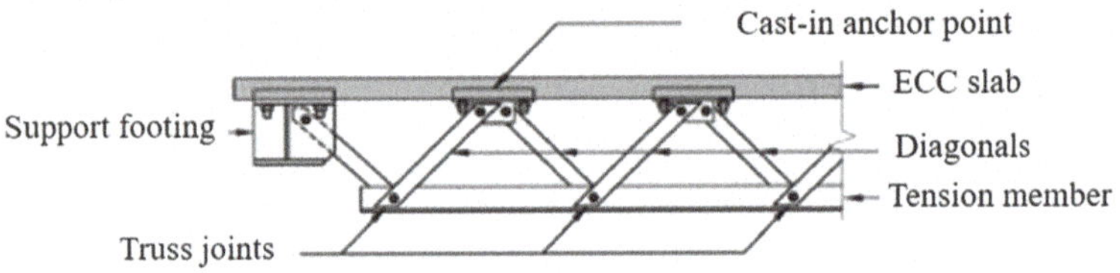

Fig. 7 Lightweight composite deck element with thin ECC slab connected to a steel truss joist [5]

suggested considering the axial loadings while checking the strength of ECC coupling beams.

3.1.2 ECCs in Modular Structures

Because of its easy installation, low cost, and excellent quality control during factory construction, modular housing has gained popularity among construction companies [31]. Fischer [81] proposed the use of ECCs to make structural components for modular structures that are durable, hazard-resilient, and lightweight. Comparing ECC panels to conventional RCC panels, they are thinner and lighter owing to their distinct tensile qualities, which eliminate the requirement for welded wire mesh or reinforcing bars. Fischer [81] examined a prefabricated polyvinyl alcohol (PVA) ECC floor panel system that offered an alternative to conventional floor systems such as wood sheathing on steel joists and RCC on panel decks, as presented in Fig. 7. With its high flexural stiffness, strength, and ductile ultimate failure, the load–deformation behavior of PVA-ECC floor panel was characterized by beneficial properties. The main benefits of this material were its excellent strength-to-weight ratio and cost. Furthermore, the eco-friendly attributes of PVA-ECC floor slab include the usage of recycled products with a minimum quantity of light gauge steel, all while preserving maximum functionality and performance. Moreover, Larusson et al. [82] assessed lightweight, prefabricated composite deck components made up of steel substructures and high-performance ECCs. Integrally, cast panels and modular panels, two panel types that demonstrated ductile failure modes and consistent load-deflection behavior, were examined, as displayed in Fig. 8. By

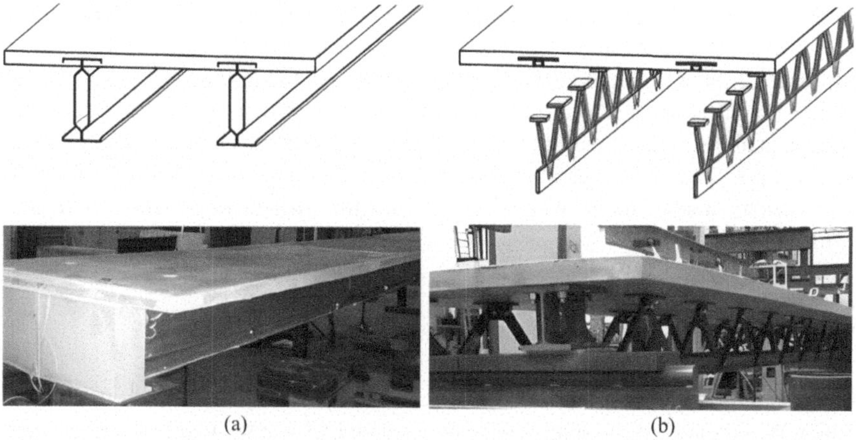

Fig. 8 Lightweight prefabricated panels: (a) integrally cast and (b) modular [82]

varying the combination of ECC deck and steel substructure, these panels provided flexibility in meeting different performance requirements. Under loads, ECC slab and associated substructure performed as a single composite element. While still meeting the fire resistance criteria, the application of ECCs improved tensile loading capacity and ductility and decreased the demand for cementitious material and transverse steel reinforcement.

3.1.3 ECC as a Retrofit Material

ECC's exceptional characteristics such as being lightweight and having self-healing properties, micro-cracking behavior, tensile strength, and ductility characteristics make it suitable for utilization in retrofitting of the existing structures [69]. Using glass fiber-reinforced polymer (FRP) sheets, adhesives, and ECCs with FA and slag, Ahmadi et al. [83] attempted to repair tensile cracking in concrete panels. The damaged panel was strengthened, while cracking and leakage were also improved by ECCs. ECCs were proved as a good alternative to RCC for repairing structures like water tanks. As compared to ECC using slag, FA-ECC had a smaller leakage rate, a lower tensile capacity, and fewer small cracks. Furthermore, the tensile bond strength of ECC's repair materials was studied by Wang et al. [84], considering roughness of surface, strength of substrate, primer, bonding agent, and curing time. Cast-in-situ ECC, precast ECC, FRC, and traditional cast-in-situ concrete were the four materials that were investigated. Up to a point, increased roughness and substrate strength increased bond strength and improved interlocking. Ensuring compatibility with ECCs, FA, SF, and expansion agents had a good impact on the primer and bonding agent, thus strengthening the bond. Besides that, curing time also increased tensile bonding strength. Huang et al. [85] examined wasted rubber in ECC's repair products, and the results exhibited better cracking resistance, limited cracks, and a reduced cracking initiation time, as listed in Table 2. Tire rubber improved the durability and tensile ductility, but because of its increased porosity, it lost some of its tensile and f'_c. Hence, they concluded that tire rubber is still useful for nonstructural applications and cannot be used for structural repairs. According to Zhou et al. [86], the excellent tensile and flexural strengths, fatigue performance, ductility, post-peak mechanical characteristics, and impact resistance of ECCs with polyethylene make them a perfect option for structural rehabilitation and fortification.

3.1.4 Application in Large-Span Structures

The structural application of ECCs has been a subject of intense study in the academia. While many of the studies have focused on ECCs at the material and element levels, there remains a gap in assessing its efficiency at the structural level. In response, the study [87] delved into the global response of ECC structures under seismic loadings, comparing their seismic performance with RCC structures. To investigate this, a large-span building structure, a 40 m span aircraft hangar, was undertaken for seismic performance assessment. Figure 9 illustrates both the architectural and structural models of this case study. Under gravity and lateral loads designed at the structural level, the members were individually designed using both conventional RCC and ECC materials. Notably, ECC members showed a reduction of 30% in longitudinal steel requirement compared to conventional RCC members, because of their enhanced tensile behavior. Moreover, ECCs had higher capacity in shear compared to RCC, therefore requiring less shear reinforcement.

Nonlinear finite element models were developed for both RCC and ECC and analyzed by pushover and nonlinear response history analyses to determine their efficiency under seismic loadings. When it comes to ductility, damage in structures, and how they behave both locally and globally, ECC did way better than RCC structures. Furthermore, the cost analysis demonstrated that the materials for ECC were about 11.9% cheaper compared to RCC. These findings highlighted how ECC could be a smart choice for getting a ductile response and improved performance during seismic events.

Table 2 Crack width, length, and numbers in ECCs at different ages

	R0 (0% recycled rubber)			R40 (40% recycled rubber)		
Age (days)	Crack width (μm)	Crack length (mm)	Crack number	Crack width (μm)	Crack length (mm)	Crack number
7	0	0	0	0	0	0
14	31	51	10	0	0	0
21	39	59	21	0	0	0
30	46	66	22	35	52	5
45	50	69	22	43	73	9
60	53	72	22	46	77	9

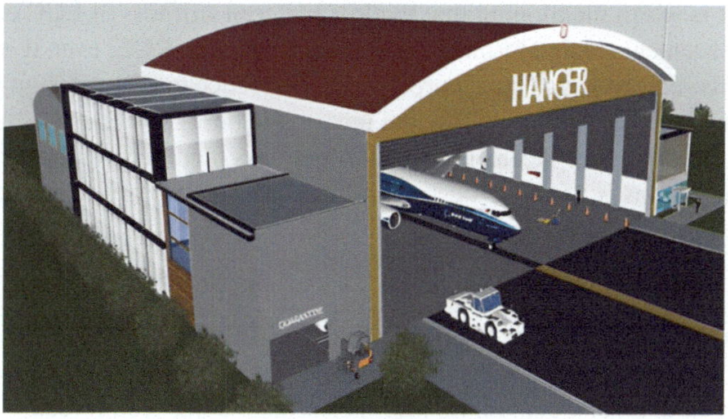

(a) Architectural model

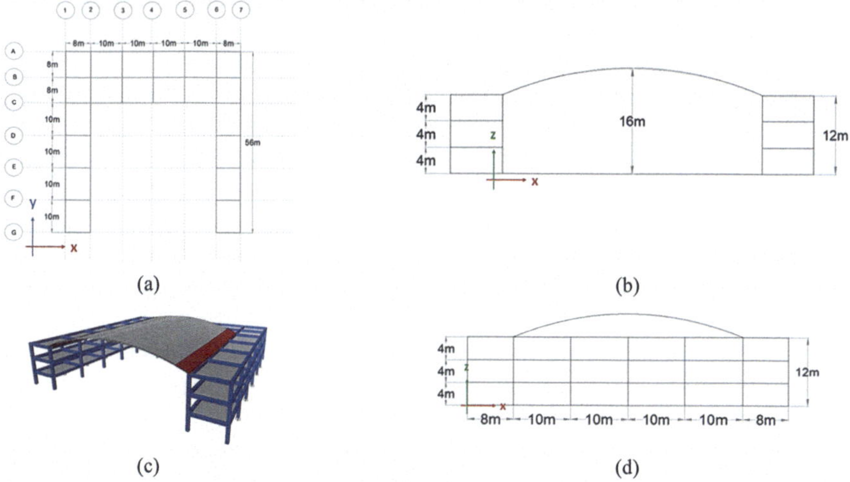

(b) Structural model

Fig. 9 Architectural and structural models of case study [88]: (**a**) architectural model and (**b**) structural model

A few important outcomes obtained from the research include the following:

- A 25% drop in structural demands was obtained owing to ECC's lesser unit weight, leading to lower dead and earthquake loads.
- Due to high tensile capacity of ECC, a 25% drop in section sizes of members (elements) was noticed, making ECC a viable substitute of RCC for large-span structures.
- More than 30% decrement in flexural reinforcement and approximately 15% drop in reinforcement required to bear compressional loads were achieved, as per JSCE [87] guidelines, validating ECC's ability to reduce steel requirement.
- Better ductility under lateral loads and less damage indicated that ECC performed seismically better than RCC.

However, it is worth mentioning that the low material stiffness of ECC can lead to higher displacements and deformations when compared to RCC, even if it is still within the limits set by codes. Thus, more research is needed to verify design parameters, check or discard the assumptions, and investigate how ECC reacts in different special structures. This also means figuring out stiffness modifiers, damping ratios, and structural responses to make design and analysis procedures more accurate and useful.

3.1.5 Application as a Passive Damping System in Building Structures

The utilization of dampers, containing high-performance fiber-reinforced cementitious composites (HPFRCCs) and steel rebars, was assessed in the study [89] to enhance the efficiency of buildings, especially in controlling seismic damage. These dampers are envisioned to participate in reducing the seismic demands of RCC buildings through performance-based engineering strategies.

Given the higher stiffness of RCC structures compared to steel structures, there is a demand for dampers with greater stiffness than conventional ones typically employed in steel structures. HPFRCC dampers offer several advantages, including selective structural performance, strength, stiffness, and ductility. These properties can be adjusted through configurations, rebar arrangements, and types of ingredients. Notably, HPFRCC dampers have compressive resistance, which is typically absent in traditional dampers, which can further improve their effectiveness in minimizing seismic damage.

Experimental results present that the utilization of HPFRCC leads to impressive shear stress capacities over 6 N/mm^2 and significant deflection capacities over 10%, associated to the resistance of shear and compression failures [89]. HPFRCC's ability to initiate multiple fine cracks significantly reduces structural damage, enabling the creation of small yet stiffer dampers capable of high energy absorption and lateral resistance. Also, Nagai et al. [90] evaluated PVA-ECC dampers, which were intended to reduce vibrations in structures. Seven small dampers and one bigger damper that was intended to fail under flexure were tested to assess their performance. The results revealed that every sample failed in flexure and had an impressive deformation capacity greater than 1/25 radian. The research further discussed the failure patterns, equivalent damping factors, and hysteresis characteristics. Finally, cyclic loading tests conducted on short-span HPFRCC dampers highlighted their efficiency in controlling failure pattern, capacity, ductility, and damage. Key results included the ductility achieved through HPFRCC, large shear stress and deflection capacities, decrement in damage of structures, and the viability of producing small, stiffer dampers with high energy absorption capacities. However, addressing the problems connected with predicting capacities and failure patterns requires further research to adjust evaluation methods and optimize performance-based design strategies. Figure 10 depicts the experimental setup of HPFRCC dampers and the installation setup of these dampers in different types of building structures.

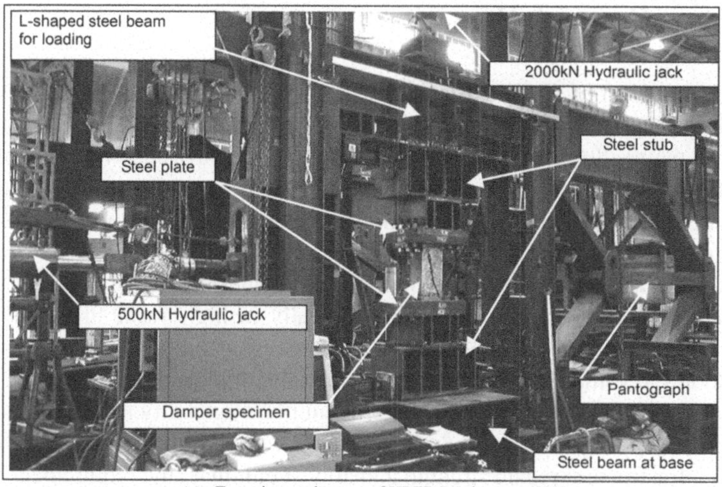

(a) Experimental setup of HPFRCC dampers [92].

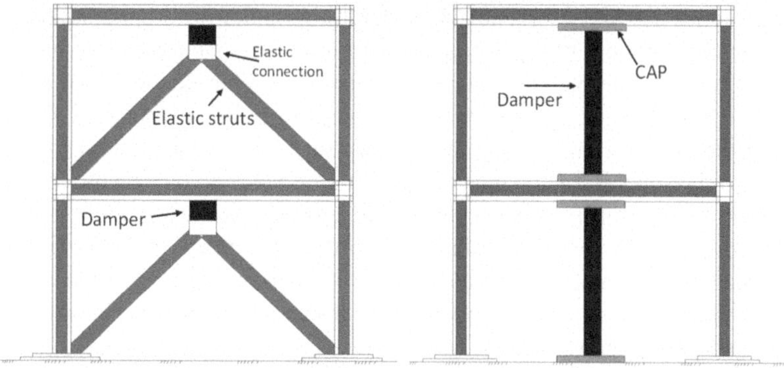

(b) Dampers with elastic chevron struts and dampers as a column.

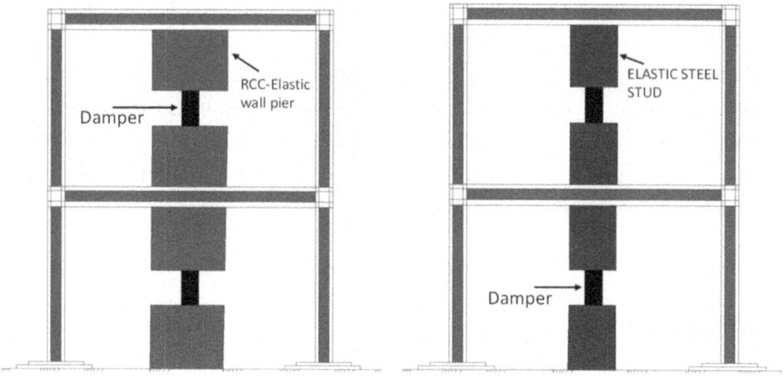

(c) Dampers with elastic RCC piers and elastic steel studs.

Fig. 10 Installation schemes for dampers in different types of building structures: (**a**) experimental setup of HPFRCC dampers [89], (**b**) dampers with elastic chevron struts and dampers as a column, and (**c**) dampers with elastic RCC piers and elastic steel studs

4 Future Research Directions

Future research on ECCs for sustainable structures shows how they can be impactful in various applications to improve durability and environmental performance. To maximize the results, their mechanical characteristics, environmental benefits, and cost-effectiveness need to be further investigated by engineers and researchers. Below are some of the new potential applications where ECCs can noticeably contribute to sustainability.

ECCs' extraordinary tensile strength and amazing ductility make them perfect for vehicular and airfield pavements, specifically in high-traffic areas and pavements subjected to high thermal stresses. ECCs can be fruitful in reducing cracking and potholes, resulting in a longer service life, and ultimately decreasing the maintenance work. Furthermore, due to its excellent bonding ability and suitability with existing concrete, ECCs are also deemed suitable for repairing purposes. It can help existing infrastructure regain strength and almost eliminate the requirement of new repairs. Also, ECCs' resistance to cracking is a beneficial property for pipe works, where avoiding leakage is crucial. Their durability under different conditions can aid in extending the life of pipes and also making them viable for sewerage pipes. ECCs can provide durable and resistant coating material, protecting against corrosion and chemical attack, thereby reducing water losses. Additionally, ECCs can be used as a shotcrete for tunnels and for the purpose of slope stabilization.

ECC overlays can recover damaged and deteriorated rigid pavements. The material's crack control ability is particularly beneficial for recovering the life of pavement overlays, as ECCs are resistant to harmful chemicals, including saltwater. Therefore, they are suitable for waterfront walls. Their energy absorption and crack resistance abilities can be fruitful in these harsh environments, improving the strength and life of the structures. Moreover, durability of ECCs and resistance to fatigue stresses make them the perfect material for railway sleepers. They offer a long-lasting alternative compared to traditionally used materials under cyclic loading while reducing the frequency of repairs. Additionally, ECCs are a good option for structures that require high energy absorption and impact resistance, like defensive bunkers, barriers, concrete armor, or buildings located in seismically active areas. Their capacity to dissipate energy aids in the absorption and reduction of applied forces, promoting sustainability and safety. These applications demonstrate how ECCs can promote more environmentally friendly construction processes, which include lesser repair and improved life.

5 Conclusions

This chapter summarizes the advancements and commercial applications of ECCs in building and transportation. ECCs' versatility promises future expansion into other domains, potentially replacing traditional construction materials like cement or concrete due to their unique characteristics such as tensile strength, ductility, durability, small crack width, and microcrack-bridging phenomenon. Also, the substitution of waste products in ECCs resulted in a nearly 61% reduction in carbon footprints. Benefits of ECCs include enhanced public safety; rapid recovery post-event; reduced installation time, cost, and labor; and lower maintenance and repair costs over the lifespan of infrastructures. For instance, ECC requires approximately 30% less longitudinal reinforcement compared to RCC. According to a few studies, no shear reinforcement is required in ECC members.

ECCs' application methods are as comparable to traditional concrete, including self-consolidating on-site casting, spraying (shotcrete), precasting, injection pumping, and hand troweling. However, the adoption of ECCs faces some challenges due to the extensive initial testing required and the construction industry's cost sensitivity. Familiarity with ECCs' benefits, such as their tensile properties, and the development of design guidelines are essential for worldwide acceptance.

Moreover, integrating material design with structural design is crucial to maximize ECCs' performance benefits, as various applications have specific performance requirements. Micromechanics should be the key tool for modifying the fibers and matrix ingredients suitable for specific ECC applications. Although ECCs' initial cost is higher than that of traditional concrete, their long-term maintenance and repair savings and a few useful applications can make them cost-effective. Furthermore, ECCs also play an important role in infrastructure sustainability and resilience, aligning with broader societal needs. Therefore, materials required for ECC mix should be made available locally to reduce the cost of this hazard-resilient composite.

References

1. B.T. Huang, K.F. Weng, J.X. Zhu, Y. Xiang, J.G. Dai, V.C. Li, Engineered/strain-hardening cementitious composites (ECC/SHCC) with an ultra-high compressive strength over 210 MPa. Compos. Commun. **26**, 100775 (2021). https://doi.org/10.1016/J.COCO.2021.100775
2. J.X. Zhu, L.Y. Xu, B.T. Huang, K.F. Weng, J.G. Dai, Recent developments in engineered/strain-hardening cementitious composites (ECC/SHCC) with high and ultra-high strength. Construct. Build Mater. **342**, 127956 (2022). https://doi.org/10.1016/J.CONBUILDMAT.2022.127956
3. V.C. Li et al., On the emergence of 3D printable engineered, strain hardening cementitious composites (ECC/SHCC). Cem. Concr. Res. **132**, 106038 (2020). https://doi.org/10.1016/J.CEMCONRES.2020.106038
4. H. Fukuyama, Y. Sato, V.C. Li, Y. Matsuzaki, H. Mihashi, Ductile engineered cementitious composite elements for seismic structural applications, in *CD Proceedings of the 12 WCEE, Paper*, vol. 1672, (2000)
5. V.C. Li, Engineered cementitious composites (ECC): bendable concrete for sustainable and resilient infrastructure, in *Engineered Cementitious Composites (ECC): Bendable Concrete for Sustainable and Resilient Infrastructure*, (2019), pp. 1–419. https://doi.org/10.1007/978-3-662-58438-5/COVER
6. B. Tahmouresi, P. Nemati, M.A. Asadi, A. Saradar, M. Mohtasham Moein, Mechanical strength and microstructure of engineered cementitious composites: a new configuration for

direct tensile strength, experimental and numerical analysis. Construct. Build Mater. **269**, 121361 (2021). https://doi.org/10.1016/J.CONBUILDMAT.2020.121361

7. S. Qian, V.C. Li, Simplified inverse method for determining the tensile strain capacity of strain hardening cementitious composites. J. Adv. Concrete Technol. **5**(2), 235–246 (2007). https://doi.org/10.3151/JACT.5.235
8. U.J. Malik et al., ANN-based predictive mimicker for the constitutive model of engineered cementitious composites (ECC). Construct. Build Mater. **420**, 135530 (2024). https://doi.org/10.1016/J.CONBUILDMAT.2024.135530
9. V.C. Li, From micromechanics to structural engineering-the design of cementitous composites for civil engineering applications. J. Struc. Mech. Earthq. Eng. **10**, 37–48 (1993)
10. U.J. Malik, F.A. Najam, S.A. Khokhar, F. Rehman, R.D. Riaz, Advancing seismic resilience: performance-based assessment of mid-rise and high-rise engineered cementitious composite (ECC) Buildings. Case Stud. Constr. Mater. **20**, e02732 (2024). https://doi.org/10.1016/J.CSCM.2023.E02732
11. A. Elsisi et al., A state-of-the-art review on structural strengthening techniques with FRPs: effectiveness, shortcomings, and future research directions. Materials **17**, 1408 (2024). https://doi.org/10.3390/MA17061408
12. Q. Zhang, V.C. Li, Development of durable spray-applied fire-resistive Engineered Cementitious Composites (SFR-ECC). Cem. Concr. Compos. **60**, 10–16 (2015). https://doi.org/10.1016/J.CEMCONCOMP.2015.03.012
13. D. Meng, T. Huang, Y.X. Zhang, C.K. Lee, Mechanical behaviour of a polyvinyl alcohol fibre reinforced engineered cementitious composite (PVA-ECC) using local ingredients. Construct. Build Mater. **141**, 259–270 (2017). https://doi.org/10.1016/J.CONBUILDMAT.2017.02.158
14. M.D. Lepech, Sustainable infrastructure systems using engineered cementitious composites, in *Structures Congress 2009: Don't Mess with Structural Engineers: Expanding Our Role*, (2009), pp. 1–10
15. T. Paulay, M.J.N. Priestley, *Seismic Design of Reinforced Concrete and Masonry Buildings*, vol 768 (Wiley, New York, 1992)
16. J. Niu, Some significant environmental issues in high-rise residential building design in urban areas. Energ. Build. **36**(12), 1259–1263 (2004). https://doi.org/10.1016/J.ENBUILD.2003.07.005
17. H.F. Campos, N.S. Klein, J. Marques Filho, Proposed mix design method for sustainable high-strength concrete using particle packing optimization. J. Clean. Prod. **265**, 121907 (2020). https://doi.org/10.1016/J.JCLEPRO.2020.121907
18. K. Liu, J. Zheng, S. Dong, W. Xie, X. Zhang, Mixture optimization of mechanical, economical, and environmental objectives for sustainable recycled aggregate concrete based on machine learning and metaheuristic algorithms. J. Build. Eng. **63**, 105570 (2023)
19. A. Witte, N. Garg, Quantifying the global warming potential of low carbon concrete mixes: comparison of existing life cycle analysis tools. Case Stud. Constr. Mater. **20**, e02832 (2024). https://doi.org/10.1016/J.CSCM.2023.E02832
20. S.A. Miller, Supplementary cementitious materials to mitigate greenhouse gas emissions from concrete: can there be too much of a good thing? J. Clean. Prod. **178**, 587–598 (2018). https://doi.org/10.1016/J.JCLEPRO.2018.01.008
21. Z. Liu et al., Carbon Monitor, a near-real-time daily dataset of global CO2 emission from fossil fuel and cement production. Sci. Data **7**(1), 392 (2020)
22. Y. Izumi, A. Iizuka, H.J. Ho, Calculation of greenhouse gas emissions for a carbon recycling system using mineral carbon capture and utilization technology in the cement industry. J. Clean. Prod. **312**, 127618 (2021). https://doi.org/10.1016/J.JCLEPRO.2021.127618
23. C.K. Chau, T.M. Leung, W.Y. Ng, A review on life cycle assessment, life cycle energy assessment and life cycle carbon emissions assessment on buildings. Appl. Energy **143**, 395–413 (2015)
24. S. Chen, Y. Teng, Y. Zhang, C.K.Y. Leung, W. Pan, Reducing embodied carbon in concrete materials: a state-of-the-art review. Resour. Conserv. Recycl. **188**, 106653 (2023)

25. M. Afzal, Y. Liu, J.C.P. Cheng, V.J.L. Gan, Reinforced concrete structural design optimization: a critical review. J. Clean. Prod. **260**, 120623 (2020). https://doi.org/10.1016/J.JCLEPRO.2020.120623
26. Z. Liu, W. Meng, Fundamental understanding of carbonation curing and durability of carbonation-cured cement-based composites: a review. J. CO2 Util. **44**, 101428 (2021). https://doi.org/10.1016/J.JCOU.2020.101428
27. E.B. El-Salakawy Brahim, Serviceability of concrete bridge deck slabs reinforced with fiber-reinforced polymer composite bars. ACI Struct. J. **101**(5), 727–736 (2004). https://doi.org/10.14359/13395
28. K.T. Soe, Y.X. Zhang, L.C. Zhang, Material properties of a new hybrid fibre-reinforced engineered cementitious composite. Construct. Build Mater. **43**, 399–407 (2013). https://doi.org/10.1016/J.CONBUILDMAT.2013.02.021
29. V.C. Li, Driving infrastructure sustainability with strain hardening cementitious composites (SHCC), in *Advances in Cement-Based Materials*, (Taylor & Francis Group, London, 2010), pp. 181–191
30. T.J. Dell, Ecc-on-simm test challenges, in *Proceedings. International Test Conference*, (IEEE, Piscataway, NJ, 1995), pp. 511–515
31. V.C. Li, *Engineered Cementitious Composites (ECC)* (2019), pp. 286–290
32. J. Sansalone et al., Retrofitting impervious urban infrastructure with green technology for rainfall-runoff restoration, indirect reuse and pollution load reduction. Environ. Pollut. **183**, 204–212 (2013)
33. G.S. Kishan, M. Sakthivel, R. Vijayakumar, N. Lingeshwaran, Life cycle assesment on tire derived fuel as alternative fuel in cement industry. Mater. Today Proc. **47**, 5483–5488 (2021)
34. T. Joensuu, R. Leino, J. Heinonen, A. Saari, Developing buildings' life cycle assessment in circular economy-comparing methods for assessing carbon footprint of reusable components. Sustain. Cities Soc. **77**, 103499 (2022)
35. C. Zhang et al., Environmental life cycle costing at the early stage for supporting cost optimization of precast concrete panel for energy renovation of existing buildings. J. Build. Eng. **35**, 102002 (2021)
36. M. Hou, Z. Li, V.C. Li, Green and durable engineered cementitious composites (GD-ECC) with recycled PE fiber, desert sand, and carbonation curing: mixture design, durability performance, and life-cycle analysis. Construct. Build Mater. **414**, 134984 (2024). https://doi.org/10.1016/J.CONBUILDMAT.2024.134984
37. K. Tuutti, Corrosion of steel in concrete. (1982)
38. P. Schissel, Corrosion of steel in concrete, in *Report of the Technical Committee 60-CSC RILEM*, (1988), p. 76
39. M. Hou, D. Zhang, V.C. Li, Material processing, microstructure, and composite properties of low carbon engineered cementitious composites (ECC). Cem. Concr. Compos. **134**, 104790 (2022)
40. X. Huang, R. Ranade, V.C. Li, Feasibility study of developing green ECC using iron ore tailings powder as cement replacement. J. Mater. Civ. Eng. **25**(7), 923–931 (2013)
41. J. Zhao, C. Yan, S. Liu, J. Zhang, S. Li, Y. Yan, Effect of solid waste ceramic on uniaxial tensile properties and thin plate bending properties of polyvinyl alcohol engineered cementitious composite. J. Clean. Prod. **268**, 122329 (2020)
42. J. Yu, G. Li, C.K.Y. Leung, Hydration and physical characteristics of ultrahigh-volume fly ash-cement systems with low water/binder ratio. Construct. Build Mater. **161**, 509–518 (2018)
43. O.A. Hisseine, A. Tagnit-Hamou, Nanocellulose for ecological nanoengineered strain-hardening cementitious composites incorporating high-volume ground-glass pozzolans. Cem. Concr. Compos. **112**, 103662 (2020)
44. Q. Wang, M.H. Lai, J. Zhang, Z. Wang, J.C.M. Ho, Greener engineered cementitious composite (ECC)–The use of pozzolanic fillers and unoiled PVA fibers. Construct. Build Mater. **247**, 118211 (2020)

45. F.B.P. Da Costa, D.P. Righi, A.G. Graeff, L.C.P. da Silva Filho, Experimental study of some durability properties of ECC with a more environmentally sustainable rice husk ash and high tenacity polypropylene fibers. Construct. Build Mater. **213**, 505–513 (2019)
46. F. Aslani, L. Wang, M. Zheng, The effect of carbon nanofibers on fresh and mechanical properties of lightweight engineered cementitious composite using hollow glass microspheres. J. Compos. Mater. **53**(17), 2447–2464 (2019)
47. J. Yu, C. Lu, C.K.Y. Leung, G. Li, Mechanical properties of green structural concrete with ultrahigh-volume fly ash. Construct. Build Mater. **147**, 510–518 (2017)
48. H.-L. Wu, D. Zhang, B.R. Ellis, V.C. Li, Development of reactive MgO-based engineered cementitious composite (ECC) through accelerated carbonation curing. Construct. Build Mater. **191**, 23–31 (2018)
49. A. Adesina, S. Das, Performance of engineered cementitious composites incorporating crumb rubber as aggregate. Construct. Build Mater. **274**, 122033 (2021). https://doi.org/10.1016/J.CONBUILDMAT.2020.122033
50. S. Ruan, J. Qiu, E.-H. Yang, C. Unluer, Fiber-reinforced reactive magnesia-based tensile strain-hardening composites. Cem. Concr. Compos. **89**, 52–61 (2018)
51. P. Zhang et al., Effects of magnesia expansive agents on the self-healing performance of microcracks in strain-hardening cement-based composites (SHCC). Mater. Today Commun. **25**, 101421 (2020)
52. T. Hanein, J.-L. Galvez-Martos, M.N. Bannerman, Carbon footprint of calcium sulfoaluminate clinker production. J. Clean. Prod. **172**, 2278–2287 (2018)
53. P. Padilla-Encinas, A. Palomo, M.T. Blanco-Varela, A. Fernández-Jiménez, Calcium sulfoaluminate clinker hydration at different alkali concentrations. Cem. Concr. Res. **138**, 106251 (2020)
54. L.H.J. Martin, F. Winnefeld, E. Tschopp, C.J. Müller, B. Lothenbach, Influence of fly ash on the hydration of calcium sulfoaluminate cement. Cem. Concr. Res. **95**, 152–163 (2017)
55. S. Park, Simulating the carbonation of calcium sulfoaluminate cement blended with supplementary cementitious materials. J. CO2 Util. **41**, 101286 (2020)
56. H. Zhu, K. Yu, V.C. Li, Citric acid influence on sprayable calcium sulfoaluminate cement-engineered cementitious composites' fresh/hardened properties. ACI Mater. J. **118**(6), 39–48 (2021)
57. H. Zhu, K. Yu, V.C. Li, Sprayable engineered cementitious composites (ECC) using calcined clay limestone cement (LC3) and PP fiber. Cem. Concr. Compos. **115**, 103868 (2021)
58. J. Yu et al., Matrix design for waterproof engineered cementitious composites (ECCs). Construct. Build Mater. **139**, 438–446 (2017)
59. S. Wu, J. Yang, R. Yang, J. Zhu, X. Sun, Self-healing acceleration and interface optimization of steel deck pavement with MgO and SAP. Construct. Build Mater. **266**, 121139 (2021)
60. J. Yu, C.K.Y. Leung, Strength improvement of strain-hardening cementitious composites with ultrahigh-volume fly ash. J. Mater. Civ. Eng. **29**(9), 05017003 (2017)
61. Y. Ling, K. Wang, W. Li, G. Shi, P. Lu, Effect of slag on the mechanical properties and bond strength of fly ash-based engineered geopolymer composites. Compos. Part B Eng. **164**, 747–757 (2019)
62. S. Zhang, V.C. Li, G. Ye, Micromechanics-guided development of a slag/fly ash-based strain-hardening geopolymer composite. Cem. Concr. Compos. **109**, 103510 (2020)
63. K. Turk, M.L. Nehdi, Flexural toughness of sustainable ECC with high-volume substitution of cement and silica sand. Construct. Build Mater. **270**, 121438 (2021)
64. R. Suryanita, H. Maizir, R. Zulapriansyah, Y. Subagiono, M.F. Arshad, The effect of silica fume admixture on the compressive strength of the cellular lightweight concrete. Results Eng. **14**, 100445 (2022)
65. F. Wang, X. Sun, Z. Tao, Z. Pan, Effect of silica fume on compressive strength of ultra-high-performance concrete made of calcium aluminate cement/fly ash based geopolymer. J. Build. Eng. **62**, 105398 (2022). https://doi.org/10.1016/J.JOBE.2022.105398

66. J. Acordi, A. Luza, D.C.N. Fabris, F. Raupp-Pereira, A. De Noni, O.R.K. Montedo, New waste-based supplementary cementitious materials: mortars and concrete formulations. Construct. Build Mater. **240**, 117877 (2020). https://doi.org/10.1016/J.CONBUILDMAT.2019.117877
67. M. Valizadeh Kiamahalleh, A. Gholampour, Y. Tang, T.D. Ngo, Incorporation of reduced graphene oxide in waste-based concrete including lead smelter slag and recycled coarse aggregate. J. Build. Eng. **88**, 109221 (2024). https://doi.org/10.1016/J.JOBE.2024.109221
68. M.A. Mosaberpanah, S.A. Umar, Utilizing rice husk ash as supplement to cementitious materials on performance of ultra high performance concrete—a review. Mater. Today Sustain. **7**, 100030 (2020)
69. H.-L. Wu, Y.-J. Du, J. Yu, Y.-L. Yang, V.C. Li, Hydraulic conductivity and self-healing performance of engineered cementitious composites exposed to acid mine drainage. Sci. Total Environ. **716**, 137095 (2020)
70. S.-H. Kang, S.-G. Hong, J. Moon, The use of rice husk ash as reactive filler in ultra-high performance concrete. Cem. Concr. Res. **115**, 389–400 (2019)
71. M. Gupta, R. Raj, A.K. Sahu, Effect of rice husk ash, silica fume & GGBFS on compressive strength of performance based concrete. Mater Today Proc. **55**, 234–239 (2022). https://doi.org/10.1016/J.MATPR.2021.06.343
72. J. Sun, Z. Zhang, G. Hou, Utilization of fly ash microsphere powder as a mineral admixture of cement: effects on early hydration and microstructure at different curing temperatures. Powder Technol. **375**, 262–270 (2020). https://doi.org/10.1016/J.POWTEC.2020.07.084
73. C.M. Martín, N.B. Scarponi, Y.A. Villagrán, D.G. Manzanal, T.M. Piqué, Pozzolanic activity quantification of hollow glass microspheres. Cem. Concr. Compos. **118**, 103981 (2021). https://doi.org/10.1016/J.CEMCONCOMP.2021.103981
74. I.A. Bukhari, R.L. Vollum, S. Ahmad, S. Juan, Shear strengthening of reinforced concrete beams with CFRP. Arab J. Sci. Eng. **62**(1), 65–77 (2010). https://doi.org/10.1680/macr.2008.62.1.65
75. X.C. Lu, J.F. Chen, L.P. Ye, J.G. Teng, J.M. Rotter, RC beams shear-strengthened with FRP: stress distributions in the FRP reinforcement. Construct. Build Mater. **23**(4), 1544–1554 (2009). https://doi.org/10.1016/j.conbuildmat.2008.09.019
76. F. Hosseini, B. Gencturk, H. Aryan, G. Cadaval, Seismic behavior of 3-D ECC beam-column connections subjected to bidirectional bending and torsion. Eng. Struct. **172**, 751–763 (2018). https://doi.org/10.1016/J.ENGSTRUCT.2018.06.054
77. T. Paulay, J.R. Binney, Diagonally reinforced coupling beams of shear walls. Int. Conc. Abstr. Portal **42**, 579–598 (1974)
78. T. Kanda, S. Nagai, M. Maruta, Y. Yamamoto, New high-rise R/C structure using ECC coupling beams, in *Second International RILEM Conference on Strain Hardening Composites*, (RILEM, Paris, 2011), pp. 289–296
79. T. Kanakubo, K. Shimizu, T. Kanda, S. Nagai, Evaluation of bending and shear capacities of HPFRCC members toward the structural application, in *Hokkaido University COE workshop on High Performance Fiber Reinforced Composites for sustainable infrastructure system*, (2007), pp. 35–44
80. Z.-Y. Zhang, R. Ding, J.-S. Fan, X. Nie, J. Zhang, Seismic performance and shear strength of coupling beams using engineered cementitious composites with different reinforcement layouts. Eng. Struct. **219**, 110895 (2020)
81. G. Fischer, Application of engineered cementitious composites (ECC) in prefabricated modular housing. ACI Spec. Publ. **268**, 17–28 (2010)
82. L.H. Lárusson, G. Fischer, J. Jönsson, Prefabricated floor panels composed of fiber reinforced concrete and a steel substructure. Eng. Struct. **46**, 104–115 (2013). https://doi.org/10.1016/J.ENGSTRUCT.2012.06.035
83. A. Ahmadi, M.R. Kianoush, M. Moslemi, M. Lachemi, H. Siad, E. Booya, Investigation on repair of tension cracks in reinforced concrete panels. Eng. Struct. **245**, 112974 (2021). https://doi.org/10.1016/J.ENGSTRUCT.2021.112974

84. B. Wang, S. Xu, F. Liu, Evaluation of tensile bonding strength between UHTCC repair materials and concrete substrate. Construct. Build Mater. **112**, 595–606 (2016). https://doi.org/10.1016/J.CONBUILDMAT.2016.02.149
85. X. Huang, R. Ranade, W. Ni, V.C. Li, On the use of recycled tire rubber to develop low E-modulus ECC for durable concrete repairs. Construct. Build Mater. **46**, 134–141 (2013). https://doi.org/10.1016/J.CONBUILDMAT.2013.04.027
86. S. Zhou, L. Xie, Y. Jia, C. Wang, Review of cementitious composites containing polyethylene fibers as repairing materials. Polymers **12**(11), 2624 (2020). https://doi.org/10.3390/POLYM12112624
87. H. Yokota, K. Rokugo, N. Sakata, JSCE recommendations for design and construction of high performance fiber reinforced cement composite with multiple fine cracks, in *High Performance Fiber Reinforced Cement Composites*, (Springer, Tokyo, 2008)
88. S.A. Khokhar, S. Memon, T. Ahmed, M.U. Basit, F.A. Najam, R.A. Khushnood, A comparative study on the seismic performance of long span engineered cementitious composite (ECC) structures. Case Stud. Constr. Mater. **20**, e03129 (2024). https://doi.org/10.1016/J.CSCM.2024.E03129
89. H. Fukuyama, H. Suwada, Experimental response of HPFRCC dampers for structural control. J. Adv. Concrete Technol. **1**(3), 317–326 (2003). https://doi.org/10.3151/JACT.1.317
90. S. Nagai, T. Kaneko, T. Kanda, M. Maruta, Structural capacity of reinforced PVA-ECC dampers, in *6th International RILEM Symposium on Fibre Reinforced Concretes*, (RILEM, Paris, 2004)

Enhancing Acid Resistance of Recycled Aggregate Concrete Using Rice Husk Ash and Hybrid Fibers

Musa Aminu Alhaji and Pushpendra Kumar Sharma

1 Introduction

Concrete is a heterogenous material consisting of different materials such as cement, aggregates, and water. Among the constituents of concrete, coarse aggregate occupies a large volume in the concrete matrix. Thus, the need for coarse aggregate is enormous due to the rapid development in urbanization and industrialization, which has resulted in the need for concrete. However, the excessive use of natural coarse aggregate (NCA) for making concrete depletes this nonrenewable natural resource, causing serious environmental issues. Extensive research has been conducted in recent years regarding the utilization of recycled concrete aggregate (RCA) in making recycled aggregate concrete (RAC) [1]. Consequently, utilizing RAC in construction can aid in mitigating the difficulties posed by huge quantities of waste concrete. Similarly, substituting NCA with RCA can help reduce the use of natural aggregates. However, the problem with this type of concrete is its poor bonding, which is caused by the presence of numerous interfacial transition zones (ITZs) within the concrete matrix that lead to poor strength and durability [2, 3]. The formation of aggregate fractures during processing weakens the aggregate and makes it more vulnerable to fluid penetration, absorption, and diffusion [4, 5]. This could be owing to weak ITZ caused by leftover cement mortar attached to RCA, making RAC more permeable than natural aggregate concrete (NAC) [3].

M. A. Alhaji
Department of Building Technology, Waziri Umaru Federal Polytechnic, Birnin Kebbi, Nigeria

School of Civil Engineering, Lovely Professional University, Phagwara, Punjab, India

P. K. Sharma (✉)
School of Civil Engineering, Lovely Professional University, Phagwara, Punjab, India
e-mail: pushpendra.23437@lpu.co.in

A. Bahrami (ed.), *Advances in Sustainable Concrete for Construction*,
https://doi.org/10.1007/978-3-031-85052-3_6

Sulfuric acid is one of the most harmful acids that attacks concrete, degrading the performance and shortening the lifespan of the concrete structures. According to Chintalapudi and Pannem [6], by oxidizing sulfide minerals, aerobic bacteria in sewage systems, industrial wastewater, and groundwater make the sulfuric acid (H_2SO_4). When calcium hydroxide (CH) interacts with sulfate ions, the main outcome of the sulfuric acid effect on concrete is gypsum. In addition, the interaction between gypsum and calcium aluminates with the matrix produces ettringite [7, 8]. Moreover, ettringite expands more than gypsum, resulting in microcracking of concrete structure due to the detrimental effect of the acid attack [9, 10]. As the reaction progresses, calcium silicate hydrate gel is decomposed via decalcification, reducing the strength of concrete [11]. Thus, incorporation of supplementary cementitious materials (SCMs) like silica fume (Sfume), bagasse ash, wood ash, fly ash, etc., as partial replacements of cement in concrete mix, can reduce the content of CH in concrete. This can minimize the severe effect of the acid attack on concrete [12]. Usually, chemical attacks on concrete cause hydration products to decompose, resulting in the deposition of SO_3 ions as gypsum [13]. This phenomenon harmed the strength of concrete. Moreover, using these by-products helps conserve nonrenewable resources while also making concrete more environmentally friendly [13].

For handling large amounts of sewage, concrete pipes are a good option both environmentally and economically. However, because of the extremely acidic environment, acid-resistant concrete is essential to prevent reinforcement corrosion. Conversely, the high permeability of RCA allows harmful agents to penetrate deeply into the matrix of RAC [14]. Therefore, when SCMs with high silica content partially replace cement, the amount of CH is reduced, which is primarily responsible for the formation of gypsum [15]. Accordingly, Koushkbaghi et al. [16] experimentally found that with the use of rice husk ash (RHA) in RAC mix, the rate of the water permeability and chloride diffusivity of RAC reduced significantly compared to the control mix. Similarly, Sasanipour et al. [17] reported that the utilization of Sfume in RAC decreased pore space within the matrix, which minimized the rate of the water absorption of RAC. The results indicated an appreciable enhancement in the durability properties of RAC. Moreover, Dimitriou et al. [18] discovered that the durability of RAC can be enhanced with the inclusion of SCMs in RAC. The study revealed that the use of Sfume in the mix showed a good resistance against the sorptivity and rapid chloride ion penetration.

Fiber reinforcement improves the modulus of rupture and splitting tensile and flexural strengths of RAC. Also, using steel fiber (SF) reduces the water absorption capacity of RAC [19, 20]. Furthermore, the bridging impact of fibers results in high-toughness concrete that can withstand impact loads [21, 22]. In addition, the bridging effect of the fiber reduces the dilapidation of concrete, owing to the detrimental consequence of the acid attack [23, 24]. However, the use of RHA and hybridization of alkali-resistant glass fiber (Ar-GF) and SF in RAC to enhance its resistance to the acid attack remains limited. Hence, this study intends to enhance the resistance of RAC to the acid attack by incorporating RHA and hybrid Ar-GF and SF. The detrimental effect of sulfuric acid attack on RAC was examined by considering the strength and weight losses after the periods of acid exposure. The characterization techniques such as field emission scanning electron microscopy (FESEM) and Fourier transform infrared spectroscopy (FTIR) were utilized to evaluate the mineralogical compositions of the concrete specimens after being exposed to sulfuric acid.

2 Research Significance

Recycling and reusing construction demolition waste are crucial for long-term construction and infrastructure management. Therefore, using RAC can reduce the need for NCA in engineering constructions while simultaneously addressing the environmental pollution occasioned by waste demolished concrete. However, when compared to NCA, RAC has inferior durability properties especially when exposed to aggressive environments. The main reason for its lower strength and durability is owing to excessive water absorption/porosity and weak ITZ of RCA, which allow penetration of acid into the concrete matrix. But previous studies revealed that SCMs such as fly ash, SF, RHA, etc. can lessen the water absorption and high porosity properties of RCA. Additionally, some investigations have demonstrated that a weak ITZ of RCA can be boosted by incorporating a certain amount of fiber. However, most previous studies have mainly been conducted on the independent effects of either Ar-GF or SF on RAC. There are limited studies that have been done on the improvement of ITZ of RCA using hybridization of Ar-GF and SF. Thus, the coupling effects of RHA with hybridization of Ar-GF and SF can enhance the compressive strength and reduce the water absorption property of RAC. However, the study examining the effects of RHA and hybridization of Ar-GF and SF on the compressive strength of RAC exposed to the sulfuric acid has not been conducted before. Therefore, this current study aimed to enhance the resistance of RAC to sulfuric acid attack by inclusion of RHA and hybridization of Ar-GF and SF.

3 Experimental Programs

3.1 Materials

3.1.1 Ordinary Portland Cement

The grade 43 ordinary Portland cement (OPC), which fulfills the IS 269:2015 requirements [25], was utilized in this research work. Table 1 displays the oxide compositions of OPC and RHA. Moreover, Fig. 1 illustrates the X-ray diffraction (XRD) pattern of OPC, and it can be noted that the major chemical compounds in OPC are in crystalline form, as evidenced by a number of sharp peaks in the

Table 1 Oxide compositions of OPC and RHA

Materials	SiO_2	Al_2O_3	CaO	Fe_2O_3	MgO	SO_2	Alkalis	LOI
OPC	21.20	5.30	65.50	4.50	–	2.60	0.90	–
RHA	91.50	0.20	0.60	0.40	0.40	0.20	3.50	3.20

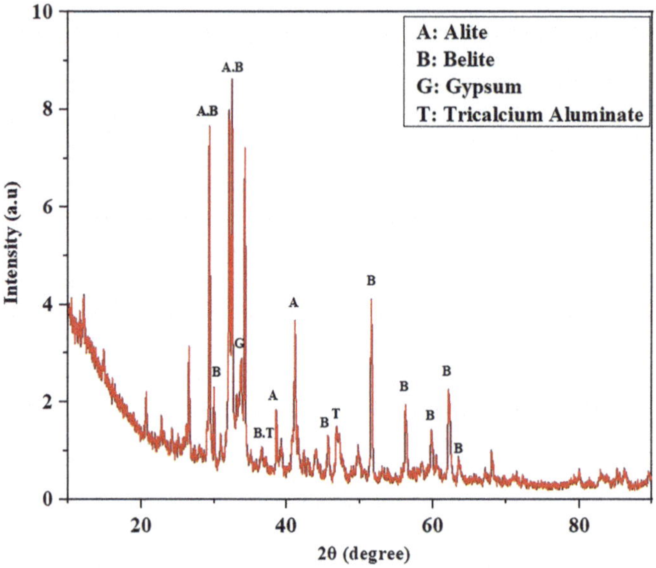

Fig. 1 XRD pattern of OPC

diffraction pattern. As it can be inferred from the graph, the main chemical constituents of OPC are alite and belite, which are primarily responsible for the strength development. Previous literature also reports a similar result for XRD pattern of OPC [26, 27]. In addition to alite and belite, the XRD result points out the presence of tricalcium aluminate and gypsum.

3.1.2 RHA

RHA was made by burning rice husk for 60 min at a temperature of roughly 750 °C in a furnace to obtain RHA with amorphous silica content of above 90% with less than 25 μm particle size. The oxide compositions of RHA presented in Table 1 indicate that it has met the requirements of ASTM C618:2005 [28] as a pozzolanic material. Furthermore, the XRD pattern of RHA depicted in Fig. 2 reveals a heightened halo diffraction pattern at 17° to 30° of 2θ range, confirming the amorphous nature of mineral phases such as SiO_2 found in RHA, which is principally responsible for the pozzolanic reaction.

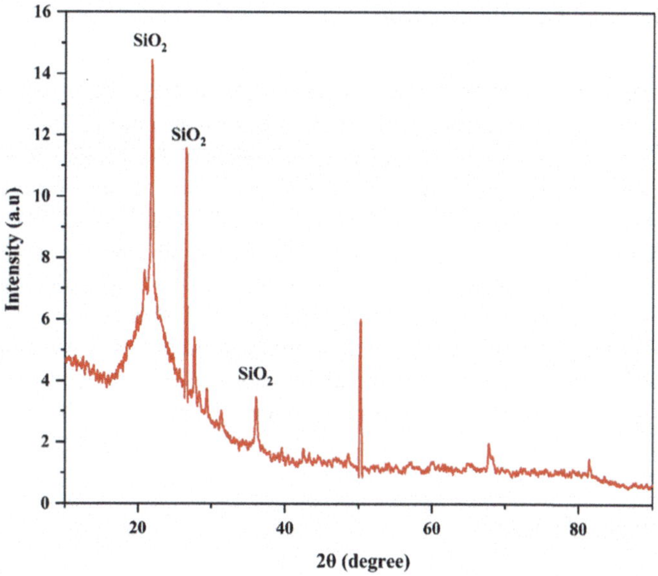

Fig. 2 XRD pattern of RHA

3.1.3 Aggregates

In the experimental work, NCA and RCA with the specific gravity values of 2.90 and 2.40, respectively, were utilized. RCA was produced by crushing the demolished concrete of grade 30 N/mm^2 obtained from Jalandhar city, Punjab. The sizes of aggregates were within the range of 60% (10 mm) and 40% (20 mm). In addition, the water absorption rates of NCA and RCA were found to be 0.60% and 3.10%, respectively. Fine aggregate (FA) employed in the experiment was sharp sand that was devoid of any extraneous substances. It fell within zone II grade and has a fineness modulus of 2.8 and a specific gravity value of 2.78.

3.1.4 Fibers

The fibers were procured from a certified dealer in Ludhiana, Punjab. The Ar-GF had a specific gravity of 2.7, a length of 30 mm, a diameter of 0.038 mm, and a tensile strength of 1400 MPa. Meanwhile, SF had a specific gravity of 7.8, and its length was 50 mm with a diameter of 1.14 mm. The tensile strength of SF ranged between 510 and 1100 MPa.

3.2 Concrete Mix Proportion

Following IS 10262:2009 [29], the concrete grade of M25 was designed in this study. NCA was partly replaced with RCA at a constant ratio of 75% in series I, while OPC was partly replaced with RHA at dosages of 10, 15, and 20%. To each of these ages, a hybridization of Ar-GF and SF in the proportions of Ar-GF0.5%-SF0.5%, Ar-GF0.25%-SF0.75%, and Ar-GF0.75%-SF0.25% by volume fraction of concrete of 1.0% was added to yield nine mix combinations in series I. In addition, the control mixes (N-concrete) with OPC100% + RHA0% + NCA100% and R-concrete mix with OPC100% + RHA0% + RCA75% (zero addition of the fibers) were cast as reference mixes for comparison to the nine mixes in series I. Thus, 11 mixtures were produced, and the mixing proportions of the materials are listed in Table 2.

3.3 Testing Methods

3.3.1 Acid Resistance Test Procedure

The detrimental effect of the sulfuric acid attack on the specimens was examined by considering the losses in the strength and weight at 28, 56, and 120 days of exposure. The hardened concrete cubes of 100 mm × 100 mm in size were utilized in the experiment. The cubes were first cured in pure water for 28 days before being exposed to 4.0% concentration solution of sulfuric acid, and weight ($W1$) of the cubes was recorded for the calculation of the weight loss, while their companion cubes were left in water for the purpose of comparison. Further, the concentration of the acid solution was checked after every 2 weeks to maintain the pH value of <1.0. The setup shown in Fig. 3 was used for the determination of the pH values every 2 weeks to measure the concentration of the sulfuric acid. This was to ensure that the pH value was maintained less than 1.0.

The age loss in the strength of the specimens was evaluated by testing the compressive strength at different test ages of 28, 56, and 120 days cured in 4.0% concentration of sulfuric acid and comparing them to water-cured specimens of the same age. The setup for the compression test is displayed in Fig. 4. The cubes were crushed at a rate of 5.15 kN/s using a HEICO compression machine of 3000 kN capacity. After the concrete cube was placed at the center of the machine, the testing of the specimens was conducted at a uniform stress rate. The specimen was loaded until it reached its maximum capacity. The ultimate compressive strength was computed by dividing the ultimate failure load by the cross-sectional area of the specimen.

Subsequently, the strength loss was determined using Eq. (1). The weight loss was determined by comparing the specimen's weight before and after exposure at different test ages, and Eq. (2) was utilized to calculate the age loss in the weight of the specimens.

Table 2 Mix proportions of materials

Group	Mix ID	Mix combinations	Materials (kg/m^3)							
			OPC	RHA	FA	NCA	RCA	Ar-GF	SF	Water
Control	N-concrete	OPC100% + RHA0% + NCA100%	438	–	618	1195	–	–	–	186
	R-concrete	OPC100% + RHA0% + RCA75%	438	–	618	299	896	–	–	186
Series I	S1–1	OPC90% + RHA10% + RCA75% + GF0.5% + SF0.5%	394.2	43.8	618	299	896	13.5	39.0	186
	S1–2	OPC90% + RHA10% + RCA75% + GF0.25% + SF0.75%	394.2	43.8	618	299	896	6.75	58.5	186
	S1–3	OPC90% + RHA10% + RCA75% + GF0.75% + SF0.25%	394.2	43.8	618	299	896	2025	19.5	186
	S1–4	OPC85% + RHA15% + RCA75% + GF0.5% + SF0.5%	372.3	65.7	618	299	896	13.5	39.0	186
	S1–5	OPC85% + RHA15% + RCA75% + GF0.25% + SF0.75%	372.3	65.7	618	299	896	6.75	58.5	186
	S1–6	OPC85% + RHA15% + RCA75% + GF0.75% + SF0.25%	372.3	65.7	618	299	896	2025	19.5	186
	S1–7	OPC80% + RHA20% + RCA75% + GF0.5% + SF0.5%	350.4	87.6	618	299	896	13.5	39.0	186
	S1–8	OPC80% + RHA20% + RCA75% + GF0.25% + SF0.75%	350.4	87.6	618	299	896	6.75	58.5	186
	S1–9	OPC80% + RHA20% + RCA75% + GF0.75% + SF0.25%	350.4	87.6	618	299	896	2025	19.5	186

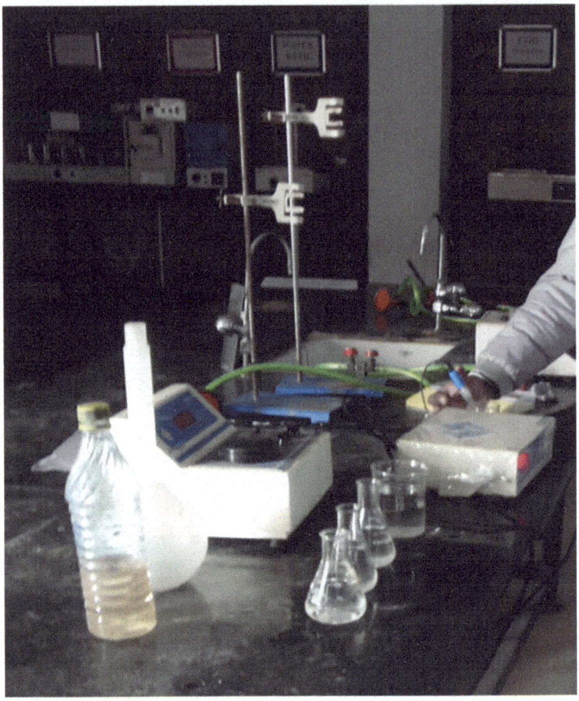

Fig. 3 Setup for testing pH values

Fig. 4 Setup for compression test

$$\text{Compressive strength loss}(\%) = \frac{\text{Change in strength after immersion in } H_2SO_4}{\text{Original strength of specimen cured in water}} \times 100\% \quad (1)$$

$$\text{Weight loss }(\%) = \frac{\text{Change in mass of specimen after immersion in } H_2SO_4}{\text{Initial mass of specimen before immersion in } H_2SO_4} \times 100\% \quad (2)$$

3.3.2 FESEM and FTIR

To study the microstructure properties, the optimum mix in series I (i.e., S1–4), the control (N-concrete), and R-concrete were selected for microstructure study at 120 days of immersion in the sulfuric acid solution. The FESEM analysis was performed on the gold-coated specimen sized 10 mm × 10 mm to study the specimens' morphological and degradation mechanisms after exposure. The JEOL brand equipment comprised of an electron gun with anode and magnetic lenses in two numbers was used in the test. For the FTIR analysis, a small sample was obtained after the compression test and grinded to a fine powder. The sample was subsequently mixed with KBr, and total attenuated reflection measurements were conducted in the frequency range 4000–500 cm^{-1}. The instrument used to do the analysis was PerkinElmer, which consisted of infrared source, interferometer, beam splitter, stationary mirror, moving mirror, and detector.

4 Results and Discussion

4.1 Compressive Strength Loss

The outcomes of loss in strength as a result of the sulfuric acid attack on various specimens after 28, 56, and 120 days of exposure are illustrated in Fig. 5. As it can be inferred from the graph, by increasing the period of immersion from 28 days to 120 days, the rate of loss in strength in all the specimens increased as well. At 28 days of exposure to 4.0% sulfuric acid, the age losses in the strength recorded for the N-concrete, R-concrete, and S1–4 (best mix in series I) specimens were 2.56%, 4.03%, and 2.2%, respectively. Furthermore, at 56 days of exposure, the age losses in the strength for the N-concrete, R-concrete, and S1–4 specimens were 5.73%, 7.57%, and 4.87%, respectively. At 120 days, the rate of losses in strength was observed to be 12.1%, 14%, and 5.77% for the N-concrete, R-concrete, and S1–4 specimens, respectively. Therefore, the R-concrete specimen had the highest age loss in terms of strength at all the periods of exposure to acid. This can be owing to the high permeability of RCA, making it more permeable to the ingress of acid

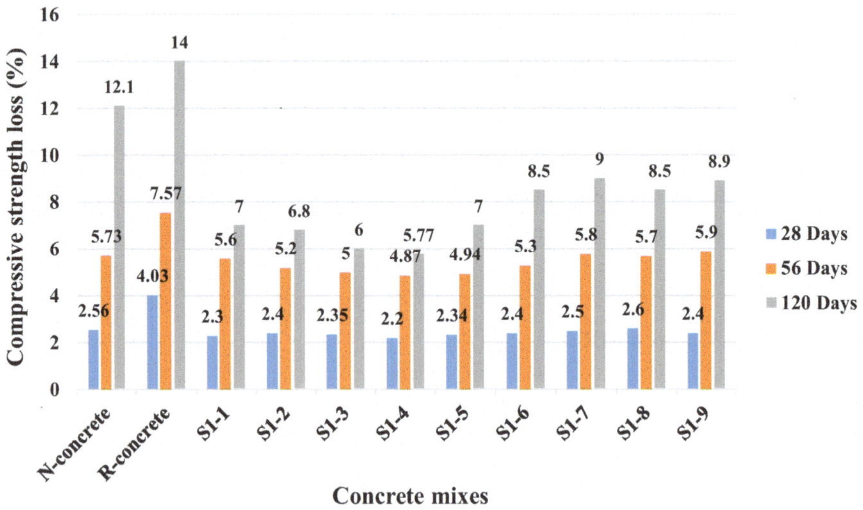

Fig. 5 Compressive strength loss due to sulfuric acid attack

solution in the matrix. This resulted in the creation of gypsum, causing the enlargement of the concrete matrix and leading to cracks within the concrete structure [30]. This caused the poor performance of R-concrete against the acid attack.

However, with the inclusion of RHA and a hybridization of Ar-GF and SF in series I, a remarkable enhancement in the resistance to the sulfuric acid attack was achieved. However, as the age replacement of RHA raised above 15, the reduction rate in the strength increased. The reason can be ascribed to the lessening of the amount of CH owing to the pozzolanic reaction with the active silica in RHA because acid reacts with CH to generate gypsum in concrete, which is the primary source of degradation of concrete. Therefore, concrete becomes less susceptible to the acid attack [16]. In addition to this, the presence of fibers has a beneficial effect on preserving the structural integrity of the binder. The incorporation of fibers has been observed to have a mitigating effect on the strength and mass deterioration caused by the acid attack. This is attributed to the ability of fibers to regulate the expansion resulting from the ettringite and gypsum formation as a result of the acid attack [23]. The incorporation of RHA with hybridization of Ar-GF and SF in RAC has been found to remarkably enhance its resistance to the acid attack. As shown in Fig. 5, the S1–4 specimen exhibited a negligible reduction in the strength among the other mixes in series I. Consequently, it can be pointed out that the S1–4 specimen demonstrated better resistance to the acid attack compared to the N-concrete and R-concrete specimens.

4.2 Weight Loss

The outcomes of the weight loss caused by the sulfuric acid attack on the specimens at the exposure periods of 28, 56, and 120 days are displayed in Fig. 6. It can be observed that as the period of immersion increased, the rate of the weight loss of the specimens increased too. However, the R-concrete specimen showed the highest weight loss compared to the other specimens. The weight loss recorded for the R-concrete specimens was 3.53%, 5.26%, and 7.26% at 28, 56, and 120 days, respectively. This was due to the numerous pores within the matrix of R-concrete, which made it less impermeable, allowing the ingress of acid in the concrete matrix and attacking CH and other hydration products [26]. In the case of the control (N-concrete) specimen, the age losses in the weight recorded after 28, 56, and 120 days of immersion were 2.51%, 3.72%, and 5.13%, respectively.

However, with the incorporation of RHA and hybrid Ar-GF and SF in series I, negligible weight losses were witnessed compared to the control (N-concrete) and R-concrete specimens. Nevertheless, as the age replacement of RHA raised above 15%, the rate of reduction in the strength also increased. Among the specimens in series I, the S1–4 specimen exhibited the most negligible reduction in terms of weight. Therefore, the age weight losses for the S1–4 specimen after the exposure ages of 28, 56, and 120 days were 2%, 3.1%, and 4%, respectively. The reason for the better performance of the S1–4 specimen over the control (N-concrete) and R-concrete specimens against the acid attack can be owing to the pozzolanic behavior of RHA, which reduced the amount of CH in concrete and made it denser, preventing the sulfuric acid solution from penetrating the specimen and lessening its degradation effects on the specimen.

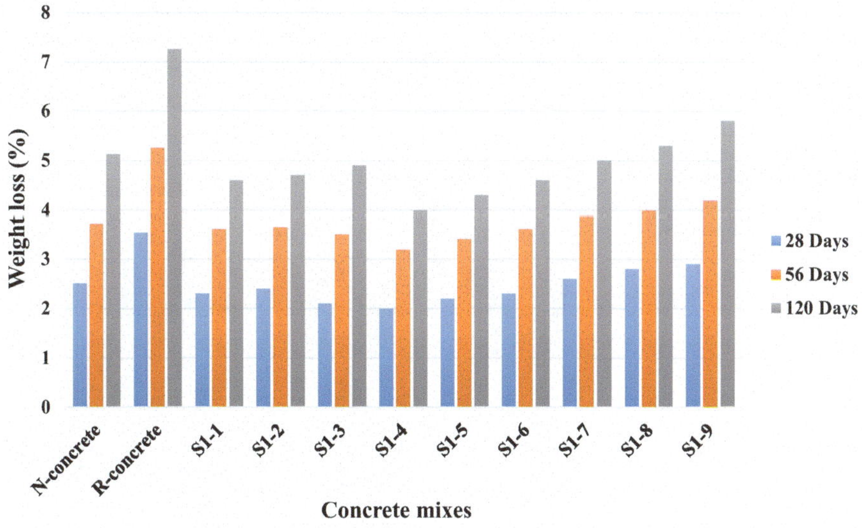

Fig. 6 Weight loss due to sulfuric acid attack

4.3 FESEM Analysis

The FESEM images of the control (N-concrete), R-concrete, and S1–4 samples at 120 days of exposure to the sulfuric acid attack are depicted in Figs. 7, 8, and 9, respectively. The FESEM images in Figs. 5 and 6 clearly demonstrate the presence of macro- and micropores in the specimens. Also, there has been a lack of particle bonding with calcium silicate hydrate (CSH) gel being decomposed due to the acid attack. The existence of prominent crystals of gypsum can be seen in the control (N-concrete) and R-concrete specimens. This can be due to the poor performance against the sulfuric acid attack of the control (N-concrete) and R-concrete specimens compared to the S1–4 specimen. Moreover, the formation of CSH gel owing to the pozzolanic reaction can be observed from the FESEM images of the S1–4 specimen, as shown in Fig. 9. This CSH gel formation reduced pore spaces within the matrix of the S1–4 specimen, which then improved its resistance to the acid

Fig. 7 FESEM image of control (N-concrete) specimen after 120 days in 4% sulfuric acid solution

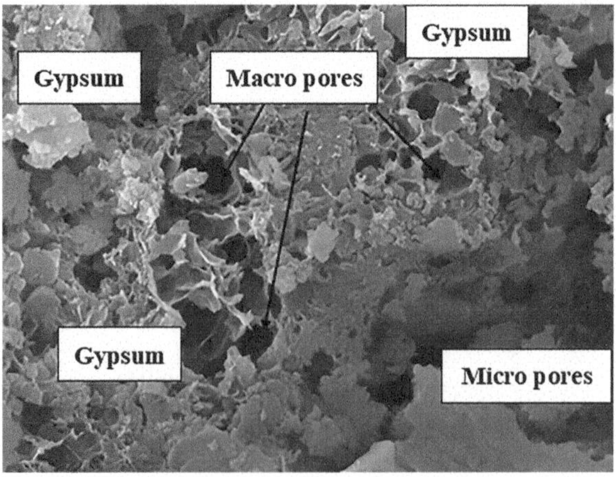

Fig. 8 FESEM image of R-concrete specimen after 120 days in 4% sulfuric acid solution

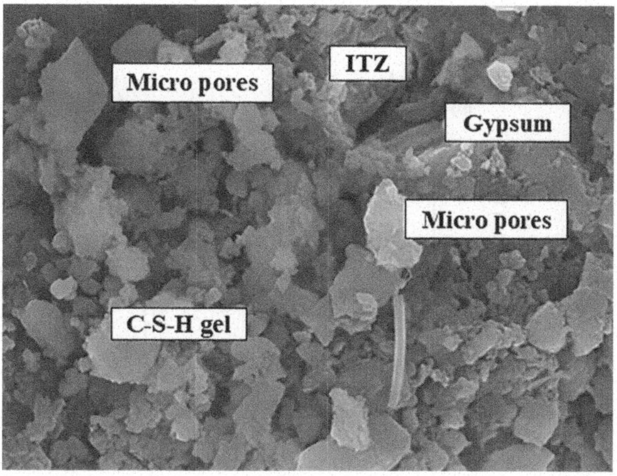

Fig. 9 FESEM image of S1–4 specimen after 120 days in 4% sulfuric acid solution

attack. The structures become more uniform and denser with the inclusion of RHA [16]. The pores were significantly reduced, which can be related to the fine particle size of RHA, which inhibited the ingress of the acid in the specimens. This issue can be because of the better performance of the S1–4 specimen against the acid attack than the control (N-concrete) and R-concrete specimens.

4.4 FTIR Analysis

The results of the FTIR analysis of the control (N-concrete), R-concrete, and S1–4 samples at 120 days of exposure to the sulfuric acid are illustrated in Figs. 10, 11, and 12, respectively. Generally, it can be observed from the graphs that the peaks at 3640–3700 cm^{-1} were due to the presence of portlandite becoming weaker, and those peaks associated with gypsum dominated the spectra in all the specimens. The peaks at 3541–3552 cm^{-1} and 3399–3402 cm^{-1}, which dominated all spectra, are attributable to the presence of gypsum. Moreover, water crystallization is responsible for the bending band at 1620 cm^{-1}, the stretching band of SO_4^{2-} at 1110 cm^{-1}, as well as the bending band at 670–602 cm^{-1} [31]. These peaks associated with gypsum suggest that the specimens were attacked by the sulfuric acid. However, these peaks corresponding to gypsum became weaker in the S1–4 specimen compared to the R-concrete and N-concrete specimens. Additionally, the asymmetric stretching of Si-O was also responsible for the broad bands at 960–980 cm^{-1}, which verifies the existence of CSH [32–34]. The intensity of these bands was reduced in the S1–4 specimen compared to the R-concrete and N-concrete specimens. This can be related to the reduction of the amount of calcium hydroxide (CH) by the active silica in RHA due to the pozzolanic reaction, which makes the S1–4 specimen less vulnerable to the acid attack.

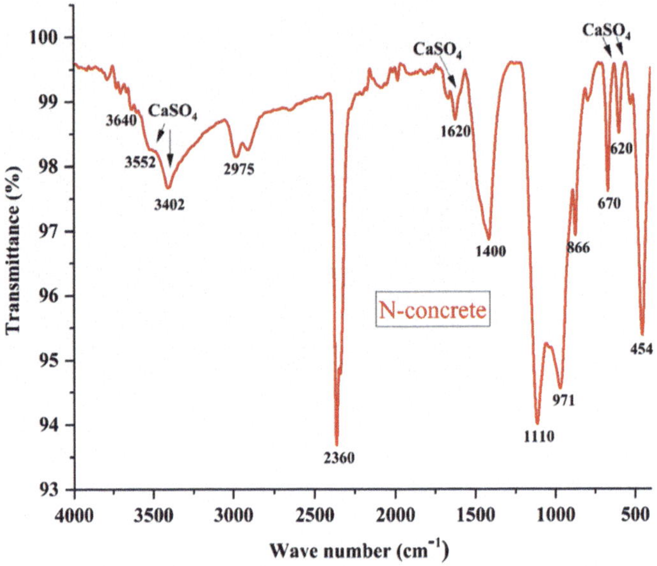

Fig. 10 FTIR of control (N-concrete) specimen at 120 days in 4% sulfuric acid solution

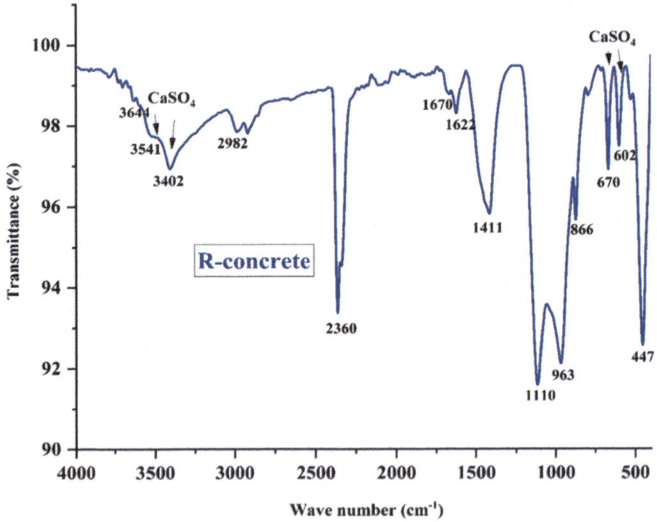

Fig. 11 FTIR of R-concrete specimen at 120 days in 4% sulfuric acid solution

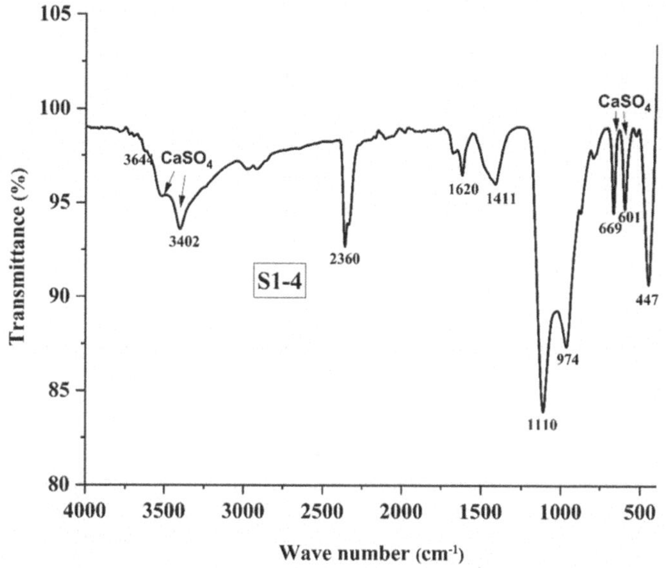

Fig. 12 FTIR of S1–4 specimen at 120 days in 4% sulfuric acid solution

5 Conclusions

This study examined methods of improving the acid resistance of RAC using RHA and hybridization of Ar-GF and SF. Based on the experimental outcomes, the following conclusions can be made:

- The use of RHA and hybridization of Ar-GF and SF in RAC effectively reduced the negative effect of acid on the specimens. This substantially enhanced the resistance of RAC to sulfuric acid.
- The S1–4 specimen exhibited a negligible loss in terms of strength and weight over the N-concrete mix (normal concrete) and all other specimens in series I. This implied that the S1–4 specimen performed better against the sulfuric acid attack than corresponding control specimens, i.e., R-concrete and N-concrete.
- The FESEM analysis conducted on the control (N-concrete), R-concrete, and S1–4 specimens at 120 days of exposure revealed the formation of pore space and crystal of gypsum within the matrix. However, micro- and macropores were less in the FESEM image of the S1–4 specimen than in the N-concrete and R-concrete specimens. This could be attributed to the reduction of the CH content by the active silica in RHA to yield a dense CSH, making the matrix dense and less permeable to the ingress of acid. Also, the incorporation of fibers was observed to have a mitigating effect on the strength and mass deterioration caused by the acid attack.
- The FTIR results demonstrated that the intensity of the peaks related to portlandite diminished, while peaks associated with the gypsum dominated the spectra,

especially for the N-concrete (control) and R-concrete specimens. This could be related to the poor performance of these specimens against the acid attack compared to the S1–4 specimen.

- The results showed that RHA is a good cement substitute in the production of RAC, especially when concreting in a chemically aggressive environment. This was as a result of strengthening the connection between concrete and fibers, which enhanced the microstructure of concrete.

Conflict of Interest None.

References

1. V. Revilla-Cuesta, M. Skaf, F. Faleschini, J.M. Manso, V. Ortega-López, Self-compacting concrete manufactured with recycled concrete aggregate: an overview. J. Clean. Prod. **262**, 121362 (2020). https://doi.org/10.1016/j.jclepro.2020.121362
2. M.A. Alhaji, P.K. Sharma, Combined effects of rice husk ash and hybrid fibres on the fresh and mechanical properties of recycled aggregate concrete. Mater. Today Proc. **62**, 6695 (2022) Accessed 2 May 2022
3. A. Sachdeva, P.K. Sharma, A. Kumar, Strength and durability properties improvement of bottom ash concrete using ultra-fine slag cementitious material. J. Mech. Eng. Res. Dev. **44**(10), 57–65 (2021) ISSN: 1024-1752
4. R. Wang, N. Yu, Y. Li, Methods for improving the microstructure of recycled concrete aggregate: a review. Construct. Build Mater. **242**(2), 118164 (2020). https://doi.org/10.1016/j.conbuildmat.2020.118164
5. M.A. Alhaji, P.K. Sharma, Studies on strength characteristic of recycled aggregate concrete using supplementary cementitious materials and hybrid fibres, in *Recent Advances in Sustainable Environment. Lecture Notes in Civil Engineering*, ed. by K.R. Reddy, S. Kalia, S. Tangellapalli, D. Prakash, vol. 285, (Springer, Singapore, 2023). https://doi.org/10.1007/978-981-19-5077-3_6
6. C. Alexandridou, G.N. Angelopoulos, F.A. Coutelieris, Mechanical and durability performance of concrete produced with recycled aggregates from Greek construction and demolition waste plants. J. Clean. Prod. **176**, 745–757 (2018). https://doi.org/10.1016/j.jclepro.2017.12.081
7. K. Chintalapudi, R.M.R. Pannem, Enhanced chemical resistance to sulphuric acid attack by reinforcing graphene oxide in ordinary and Portland Pozzolana cement mortars. Case Stud. Constr. Mater. **17**, e01452 (2022). https://doi.org/10.1016/j.cscm.2022.e01452
8. T. Gutberlet, H. Hilbig, R.E. Beddoe, Acid attack on hydrated cement—effect of mineral acids on the degradation process. Cem. Concr. Res. **74**, 35–43 (2015). https://doi.org/10.1016/j.cemconres.2015.03.011
9. M. Muthu, E.H. Yang, C. Unluer, Effect of graphene oxide on the deterioration of cement pastes exposed to citric and sulfuric acids. Cem. Concr. Compos. **124**, 104252 (2021)
10. V. Zivica, A. Bajza, Acidic attack of cement-based materials—a review. Part 1. Principle of acidic attack. Construct. Build Mater. **15**, 331–340 (2001). https://doi.org/10.1016/S0950-0618(01)00012-5
11. V. Zivica, A. Bajza, Acidic attack of cement-based materials—a review part 2. Factors of rate of acidic attack and protective measures. Construct. Build Mater. **16**, 215–222 (2002)
12. I.K. Jeon, A. Qudoos, S.H. Jakhrani, H.G. Kim, J.S. Ryou, Investigation of sulfuric acid attack upon cement mortars containing silicon carbide powder. Powder Technol. **359**, 181–189 (2020). https://doi.org/10.1016/j.powtec.2019.10.026

13. Z.T. Chang, X.J. Song, R. Munn, M. Marosszeky, Using limestone aggregates and different cements for enhancing resistance of concrete to sulphuric acid attack. Cem. Concr. Res. **35**, 1486–1494 (2005). https://doi.org/10.1016/j.cemconres.2005.03.006
14. D.A. Adesanya, A.A. Raheem, A study of the permeability and acid attack of corn cob ash blended cements. Construct. Build Mater. **24**, 403–409 (2010). https://doi.org/10.1016/j.conbuildmat.2009.02.001
15. G. Yakovlev, И. Полянских, A. Gordina, et al., Influence of sulphate attack on properties of modified cement composites. Appl. Sci. **11**, 8509 (2021). https://doi.org/10.3390/app11188509
16. B. Masood, A. Elahi, S. Barbhuiya, B. Ali, Mechanical and durability performance of recycled aggregate concrete incorporating low calcium bentonite. Construct. Build Mater. **237**, 117760 (2020). https://doi.org/10.1016/j.conbuildmat.2019.117760
17. M. Koushkbaghi, M.J. Kazemi, H. Mosavi, E. Mohseni, Acid resistance and durability properties of steel fiber-reinforced concrete incorporating rice husk ash and recycled aggregate. Construct. Build Mater. **202**, 266–275 (2019)
18. H. Sasanipour, F. Aslani, J. Taherinezhad, Effect of silica fume on durability of self-compacting concrete made with waste recycled concrete aggregates. Construct. Build Mater. **227**, 116598 (2019)
19. G. Dimitriou, P. Savva, M.F. Petrou, Enhancing mechanical and durability properties of recycled aggregate concrete. Construct. Build Mater. **158**, 228–235 (2018). https://doi.org/10.1016/j.conbuildmat.2017.09.137
20. V. Afroughsabet, L. Biolzi, T. Ozbakkaloglu, Influence of double hooked-end steel fibers and slag on mechanical and durability properties of high performance recycled aggregate concrete. Compos. Struct. **181**, 273–284 (2017)
21. B. Ali, L.A. Qureshi, S.U. Khan, Flexural behavior of glass fiber-reinforced recycled aggregate concrete and its impact on the cost and carbon footprint of concrete pavement. Construct. Build Mater. **262**, 120820 (2020). https://doi.org/10.1016/j.conbuildmat.2020.120820
22. A.B. Kizilkanat, N. Kabay, V. Akyüncü, S. Chowdhury, A.H. Akça, Mechanical properties and fracture behavior of basalt and glass fiber reinforced concrete: An experimental study. Construct. Build Mater. **100**, 218–224 (2015)
23. B. Ali, L.A. Qureshi, R. Kurda, Environmental and economic benefits of steel, glass, and polypropylene fiber reinforced cement composite application in jointed plain concrete pavement. Compos. Commun. **22**, 100437 (2020). https://doi.org/10.1016/j.coco.2020.100437
24. B. Ali, L.A. Qureshi, S.H.A. Shah, S.U. Rehman, I. Hussain, M. Iqbal, A step towards durable, ductile and sustainable concrete: simultaneous incorporation of recycled aggregates, glass fiber and fly ash. Construct. Build Mater. **251**, 118980 (2020). https://doi.org/10.1016/j.conbuildmat.2020.118980
25. D. Gao, L. Zhang, J. Zhao, P. You, Durability of steel fibre-reinforced recycled coarse aggregate concrete. Construct. Build Mater. **232**, 117119 (2020). https://doi.org/10.1016/j.conbuildmat.2019.117119
26. Indian Standard IS 269, *Specification of Requirements of Ordinary Portland cement* (Bureau of Indian Standards, New Delhi, 2015)
27. R. Muduli, B.B. Mukharjee, Performance assessment of concrete incorporating recycled coarse aggregates and metakaolin: a systematic approach. Construct. Build Mater. **233**, 117223 (2020). https://doi.org/10.1016/j.conbuildmat.2019.117223
28. R. Singh, D. Nayak, A. Pandey, R. Kumar, V. Kumar, Effects of recycled fine aggregates on properties of concrete containing natural or recycled coarse aggregates: a comparative study. J. Build. Eng. **45**, 103442 (2022). https://doi.org/10.1016/j.jobe.2021.103442
29. American Society for Testing and Material, ASTM C618–05, *Specification for Fly Ash and Raw or Calcined Natural Pozzolana or Use in as a Mineral Admixture in Portland cements Concrete* (American Society for Testing and Material, West Conshohocken, 2005)
30. Indian Standard IS: 10262, *Proportioning-Guidelines. Bureau of Indian Standards* (Bureau of Indian Standards, New Delhi, 2009)

31. B. Chatveera, P. Lertwattanaruk, Durability of conventional concretes containing black rice husk ash. J. Environ. Manage. **92**, 56–66 (2011). https://doi.org/10.1016/j.jenvman.2010.08.007
32. J. Usman, A.R.M. Sam, Acid resistance of palm oil fuel ash and metakaolin ternary blend cement mortar. Sustain. Environ. Res. **27**, 181–187 (2017). https://doi.org/10.1016/j.serj.2017.02.003
33. I.G. Lodeiro, D.E. Macphee, A. Palomo, A. Fernández-Jiménez, Effect of alkalis on fresh C–S–H gels FTIR analysis. Cem. Concr. Res. **39**, 147–153 (2009). https://doi.org/10.1016/j.cemconres.2009.01.003
34. V.S. Kashyap, U. Agrawal, K. Arora, G. Sancheti, FTIR analysis of nanomodified cement concrete incorporating nano silica and waste marble dust. IOP Conf. Ser. Earth Environ. Sci. **796**, 012022 (2021). https://doi.org/10.1088/1755-1315/796/1/012022

Internal Curing Concrete: A New Age Construction Technique

V. Kastro Kiran and Dhanya Sathyan

1 Introduction

Curing of concrete is essential for achieving long-lasting and functional structures as it allows for complete cement hydration, leading to increased strength and minimized cracks, particularly shrinkage cracks. These cracks not only weaken the structure but also accelerate the deterioration due to exposure and environmental factors. Proper curing also reduces water permeability, enhancing resistance to freeze–thaw damage, deicing salts, and harmful substances. Curing promotes the formation of a denser and less porous concrete matrix, offering several crucial benefits to concrete. It reduces the permeability of concrete, making it more resistant to water penetration, chloride ingress (which can cause corrosion of steel reinforcement), and deicing salts used on roads in cold climates. Additionally, a well-cured concrete exhibits improved freeze–thaw resistance.

In addition to its immediate effects on strength development and crack prevention, concrete curing also contributes remarkably to long-term durability of structures. Proper curing helps mitigate the risk of distress mechanisms such as alkali–silica reaction (ASR) and sulfate attack by providing a dense and impermeable concrete matrix. Increasing the density of concrete improves its ability to withstand harsh chemical conditions and prevents corrosion of reinforcement inside, ultimately lengthening the lifespan of the building. Moreover, adequate curing assists in preserving dimensional integrity and lessening the chance of distortion or misshaping.

V. Kastro Kiran · D. Sathyan (✉)
Department of Civil Engineering, Amrita School of Engineering, Amrita Vishwa Vidyapeetham, Coimbatore, India
e-mail: s_dhanya@cb.amrita.edu

A. Bahrami (ed.), *Advances in Sustainable Concrete for Construction*,
https://doi.org/10.1007/978-3-031-85052-3_7

1.1 Difficulties Associated with Traditional Curing Techniques

Conventional concrete curing methods, like ponding, spraying, or covering with wet bags, can be cumbersome and inefficient. Insufficient moisture retention can lead to rapid surface drying, causing plastic shrinkage cracking and hindering strength development. These methods often consume substantial amounts of water, making them unsustainable, particularly in water-stressed areas. Furthermore, logistical limitations, such as limited access to curing facilities or unfavorable weather conditions, can further complicate the curing process, potentially compromising the final quality and durability of concrete. In arid climates, the high evaporation rate necessitates frequent reapplication of water to prevent premature drying, which can be labor-intensive and impractical, especially for large-scale concrete pours.

Internally cured concrete (ICC) provides a groundbreaking solution to the challenges posed by traditional curing methods. Integrating internal curing agents directly into the concrete mix offers a sustainable and efficient approach to moisture management. These agents function as microscopic water reservoirs, gradually releasing moisture as the cement hydrates. This internal moisture supply simplifies the curing process, ensuring a more uniform and consistent hydration reaction. As a result, ICC delivers improved strength, reduced shrinkage cracking, and enhanced overall durability.

1.2 Significance of Study

This chapter holds significant potential to revolutionize concrete curing practices by providing a comprehensive understanding of internal curing mechanisms and their impact on concrete performance. By unraveling the role of internal curing agents, such as polyethylene glycol (PEG) and superplasticizers, in enhancing the hydration process, this chapter offers valuable insights for optimizing concrete mix designs to achieve superior mechanical properties and durability. The findings are particularly relevant in addressing the challenges associated with traditional curing methods, including water scarcity and environmental concerns. Furthermore, this research contributes to the advancement of concrete technology by proposing innovative solutions to improve the long-term durability and performance of concrete structures, ultimately extending their service life and reducing maintenance costs. The adoption of internal curing techniques, as detailed in this study, can lead to the development of more resilient and sustainable infrastructure, addressing critical issues in contemporary construction practices.

Fig. 1 Materials used for ICC

2 IC: A Multifaceted Approach

The American Concrete Institute (ACI) defines IC as "a process by which the hydration of cement continues because of the availability of internal water that is not part of the mixing water" [1]. This essentially means supplying water for cement hydration from within the concrete itself, rather than relying solely on external measures like surface wetting. This technique involves incorporating lightweight, moisture-absorbing aggregates or pre-saturated lightweight aggregates (LWAs) into the concrete mixture. These aggregates release stored water during the curing process, helping maintain adequate moisture levels within concrete, thus improving hydration and reducing the risk of cracking due to shrinkage. IC can enhance the durability and performance of concrete structures, particularly in environments where drying shrinkage and cracking are significant concerns. Figure 1 shows the various materials that can be used to prepare ICC.

2.1 *Importance of IC in Concrete Construction*

In the realm of concrete construction, unlike traditional methods that rely solely on external water sources, IC provides an ingenious solution by incorporating a built-in moisture reservoir within the concrete mix. This internal water supply plays a pivotal role in mitigating shrinkage and cracking, ultimately leading to a more durable and robust concrete structure. By promoting a more complete hydration process, IC can contribute to a slight increase in compressive strength in certain concrete mixes. Moreover, the formation of a denser concrete matrix with fewer voids results in reduced permeability. This improved impermeability protects concrete from harmful factors like water infiltration and chloride ingress, thereby enhancing its overall durability.

IC offers a distinct advantage over traditional external curing methods. Conventional methods like ponding or using wet burlap are often labor-intensive and inefficient due to water evaporation. Maintaining consistent external curing, particularly in adverse weather conditions, can be challenging. IC provides a more reliable and sustainable approach by ensuring consistent moisture availability within concrete [2]. From a sustainability perspective, IC enhances resource efficiency and supports environmental preservation in concrete construction. It reduces the need for external water during the curing process and minimizes the potential for future concrete repairs or replacements, thus saving both water and materials. In addition, by prolonging the lifespan of concrete structures, IC aligns with sustainable development goals, promoting long-term resource conservation and lowering overall life-cycle costs.

2.2 Historical Background and Evolution of IC Technique

The groundwork for IC can be traced back to the observations of T.C. Powers and colleagues in the 1940s. Their research identified the potential benefits of internally supplied water for continued hydration within concrete. Despite these early observations, widespread adoption of IC remained limited in the mid-twentieth century. Traditional external curing methods like water curing or wet burlap continued to be the dominant practices. The use of LWAs in concrete dates back to the mid-twentieth century. Initially, LWAs such as expanded clay, shale, and slate were primarily utilized to reduce the density of concrete and improve its thermal properties. However, researchers soon recognized their potential for IC due to their ability to absorb and release moisture. By the late twentieth century, the construction industry began to recognize the limitations of external curing, particularly regarding consistency and resource intensiveness. Further, the development of high-performance concretes with slower hydration rates due to the use of supplementary cementitious materials (SCMs) like fly ash or slag highlighted the need for more reliable hydration methods. With increasing awareness of the benefits of IC, its adoption in concrete construction has been steadily growing. Today, IC techniques are widely used in various applications, including high-performance concrete, bridge decks, pavements, and other critical infrastructure projects where durability and longevity are paramount. Its advantages in terms of improved performance, reduced environmental impact through lower water consumption, and cost-effectiveness compared to some external curing methods are driving its wider acceptance. As research on IC progressed, industry organizations such as ACI started recognizing its potential benefits and began incorporating provisions related to IC into their codes and standards. ACI 308R-16 includes provisions and recommendations for IC techniques.

2.3 Introduction to Three Primary IC Agents

The introduction of the three primary IC agents, namely PEG, superabsorbent polymers (SAPs), and LWAs, marks noticeable advancements in concrete technology aimed at improving hydration, reducing shrinkage, and enhancing durability.

2.3.1 PEG

PEG is a water-soluble polymer commonly used as an internal curing agent in concrete. PEG molecules have the capacity to absorb and retain substantial amounts of water, which they can then release gradually during the curing process. When added to concrete mixtures, PEG acts as a reservoir of moisture, ensuring continuous hydration of cement and promoting uniform water distribution within the concrete matrix [3]. This helps reduce shrinkage and cracking, leading to improved durability and performance of concrete structures.

2.3.2 SAPs

SAPs are synthetic materials with an extraordinary capacity to absorb water. These polymers can absorb hundreds of times their weight in water, forming gel-like structures that retain moisture within the concrete matrix. When added to concrete mixtures, SAP particles absorb excess mixing water and release it gradually during hydration, providing internal curing and maintaining optimal moisture levels. By preventing moisture loss and reducing shrinkage, SAPs enhance strength, durability, and service life of concrete structures, especially in harsh environmental conditions [4].

2.3.3 LWAs

LWAs, such as expanded clay, shale, or slate, are porous materials with high moisture absorption capacity. These aggregates are typically presoaked or pre-wetted before being incorporated into concrete mixtures [5]. During the curing process, moisture stored within the porous structure of LWAs is gradually released, ensuring continuous hydration of cementitious materials and IC of concrete. LWAs help minimize shrinkage, improve durability, and enhance the performance of concrete structures, particularly in applications where weight reduction is desired or where

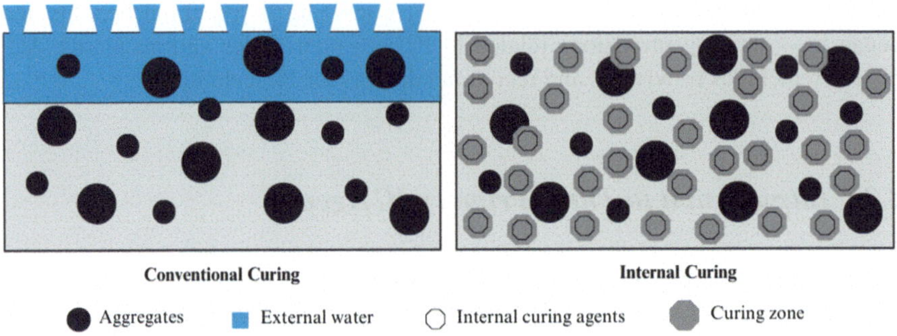

Fig. 2 Mechanism of conventional curing and IC in concrete

conventional curing methods may be impractical. The mechanism of IC in concrete supplied by using IC agents is illustrated in Fig. 2.

3 PEG as an IC Agent

PEG emerges as a pioneering IC agent in concrete technology, offering unique capabilities to address hydration challenges and enhance the performance of concrete structures. PEG plays a crucial role in minimizing early-age evaporation and promoting hydration by providing a consistent source of moisture within the concrete matrix [6]. During the initial stages of hydration, when concrete is most susceptible to moisture loss due to evaporation, PEG serves as a protective barrier, preventing rapid drying of the concrete surface. By retaining moisture within the pore structure, PEG reduces the rate of evaporation and maintains favorable curing conditions, even in adverse environments with high temperatures or low humidity [7]. This ensures that the hydration process proceeds uninterrupted, allowing cementitious materials to fully react and develop the desired strength and durability properties. As a result, concrete structures incorporating PEG as an IC agent exhibit enhanced performance characteristics, including increased strength, reduced shrinkage, and improved resistance to cracking and deterioration. The typical properties of PEGs are listed in Table 1.

3.1 Mechanism of Moisture Supply by PEG in Concrete

Unlike some IC methods that rely on physical absorption, PEG employs a clever approach, that is, water sequestration. Through chemical interaction, PEG molecules form complexes with water molecules, effectively trapping moisture within the pore structure of concrete. This retained water is gradually released over time, driven by the equilibrium between the PEG–water complexes and the surrounding

Table 1 Typical properties of PEGs

S. no.	Property	Description	Typical values/ranges
1	Molecular weight	The average mass of PEG molecules. Higher-molecular-weight PEGs have better water retention	200–20,000 g/mol
2	Water holding capacity	The ability to absorb and retain moisture, providing sustained IC in concrete	20–40% by weight (depends on molecular weight)
3	Solubility	The ability of PEG to dissolve in water, ensuring uniform distribution in concrete	Highly soluble (up to 50% w/v)
4	Compatibility with concrete mix	The ability of PEG to blend with cement, aggregates, and admixtures without adverse effects	High compatibility with standard concrete components
5	Viscosity	The measure of PEG's resistance to flow, affecting its diffusion in concrete	Low to high viscosity (dependent on molecular weight)

environment [8]. The gradual release of moisture ensures a continuous supply of hydration water to cementitious materials, promoting uniform hydration and minimizing the risk of early-age shrinkage and cracking. By maintaining optimal moisture levels throughout the concrete matrix, PEG facilitates the formation of denser and more durable concrete, enhancing its long-term performance and resilience. PEG in the market is available in the forms of solution and pellets with molecular weights starting from 200 to 10,000.

3.2 *Incorporation Techniques and Dosage of PEG in Concrete Mixtures*

PEG presents a promising approach to IC in concrete. However, incorporating it effectively requires careful consideration of both technique and dosage. The most common method involves pre-dissolving PEG in a portion of the mixing water before it is added to the concrete mix. This ensures even distribution throughout the concrete and minimizes the risk of clumping. In some cases, PEG can be dry blended with the other dry ingredients of the concrete mix. However, this method requires a more precise particle size distribution for PEG to ensure proper dispersion and avoid potential issues with workability. Several manufacturers offer pre-blended concrete admixtures containing PEG. These can simplify the incorporation process and guarantee consistent quality.

The selection of PEG with different molecular weights allows for tailored water release profiles. Lower-molecular-weight PEGs release water faster, while higher weights offer a more sustained release. Different molecular weights of PEG have varying water holding capacities. Lower-molecular-weight PEGs typically require a higher dosage to achieve the same level of IC compared to higher-molecular-weight ones. This flexibility allows for optimizing the IC process based on specific concrete mix designs and curing requirements. The amount of used PEG depends on the targeted level of IC needed for the specific project requirements [9]. Factors like climate, concrete mix design, and desired level of shrinkage reduction influence the optimal dosage. Using too little might not provide sufficient internal moisture, while an excess can potentially impact the setting time, bleeding, and workability of concrete.

3.3 Impact of PEG on Fresh and Hardened Concrete Properties

PEG can affect workability of fresh concrete by modifying its rheological properties. When incorporated into concrete mixtures, PEG may act as a lubricant, reducing friction between particles and improving flowability. This can lead to increased workability and ease of placement, allowing for better consolidation and reduced voids. Lower dosages of PEG might have a minimal impact on workability, or even lead to a slight improvement due to lubrication effects [10]. Higher dosages, however, can lead to a decrease in workability. This is because the PEG molecules can occupy some of the water needed for hydration, reducing the available water for lubrication and potentially increasing the water demand of the mix.

The addition of PEG may alter the concrete setting times based on dosage and type [11]. PEG can extend the setting time by delaying the hydration process, especially in cases where high dosages are used. This is because PEG can temporarily restrict the interaction of water with cement particles, delaying the initial hydration process. Lower dosages may have minimal impact on setting time. However, proper dosage control is necessary to avoid excessive delays in setting time, which could impact construction schedules. PEG can reduce segregation and bleeding in fresh concrete by improving particle dispersion and enhancing cohesion. This results in more uniform distribution of aggregates and reduced tendency for water to migrate to the surface, leading to improved homogeneity and surface finish.

PEG can have a large influence on the early-age and long-term strength development of concrete. The effect of PEG on compressive strength can be positive, particularly at optimal dosages. The sustained availability of internal water from PEG can promote more complete hydration, leading to a denser and stronger concrete matrix. Studies have demonstrated that using PEG within an appropriate dosage range can result in an increase in compressive strength compared to conventionally cured concrete.

PEG can enhance the durability of concrete by improving its resistance to various deteriorative mechanisms, including freeze–thaw cycles, chemical attack, and ASR [12]. By promoting a denser microstructure and reducing shrinkage cracking, PEG can contribute to improved concrete durability. The continuous hydration facilitated by PEG promotes the formation of denser, less permeable concrete with increased resistance to penetration by deleterious substances. This results in longer service life and reduced maintenance requirements for concrete structures. One of the primary benefits of PEG is its ability to mitigate shrinkage cracking. By providing an internal water source, PEG mitigates autogenous shrinkage by allowing for continued hydration and reducing the overall voids within concrete [7]. This leads to improved dimensional stability and enhanced durability of concrete structures, particularly in environments prone to moisture loss. Additionally, PEG can minimize drying shrinkage to some extent by reducing the rate of moisture loss from concrete.

3.4 Applications of PEG for IC

PEG is particularly beneficial for precast concrete elements where external curing practices like water curing might be impractical or challenging to implement consistently. The internal water reservoir provided by PEG ensures continued hydration during the critical early stages, leading to improved strength and reduced shrinkage cracking. Modern high-performance concrete (HPC) mixes often incorporate SCMs like fly ash or slag. While these can enhance strength and sustainability, they can also provide slower hydration rates. The PEG's ability to provide sustained internal moisture becomes crucial for optimizing hydration and achieving desired performance characteristics in HPC. In hot weather conditions, rapid evaporation of external moisture can be detrimental to proper hydration [13]. PEG acts as a buffer, mitigating the impact of early-age drying and ensuring continued hydration for a longer duration. Concrete elements with a high surface-area-to-volume ratio are more susceptible to shrinkage cracking due to faster moisture loss. PEG's internal water reservoir helps minimize shrinkage in such applications.

3.5 Limitations of Using PEG in Concrete Construction

While PEG offers a compelling solution for IC in concrete, it is not without limitations. The limitations of PEG are as follows:

Dosage Optimization: Finding the optimal dosage of PEG is crucial. If insufficient PEG is used, it might not provide enough internal moisture to effectively mitigate shrinkage cracking, potentially defeating the purpose of IC. Using an excessive amount of PEG can also have negative consequences on setting time and strength. Excess PEG might disrupt the normal distribution of water within the concrete mix, leading to bleeding or segregation.

Compatibility Considerations: While generally compatible with many admixtures, it is essential to ensure compatibility with all the components used in the concrete mix. Consulting with admixture manufacturers and conducting compatibility tests are recommended before finalizing the PEG dosage.

Cost-effectiveness: Compared to some other IC methods, PEG can be a cost-effective option. However, the specific cost depends on the chosen dosage and the prevailing market price of PEG.

Environmental Impact: While offering benefits in terms of reduced water consumption during curing, the production process for some types of PEGs might have an environmental footprint. It is important to consider the life cycle impact of the chosen PEG product. Additionally, the disposal of PEG-containing concrete waste should be managed responsibly to minimize environmental impact.

Limited Research: While research on PEG as an IC agent is ongoing, it is a relatively new technology compared to traditional methods. More long-term performance data and studies on the interaction of PEG with different concrete mix designs are needed to fully understand its long-term effects.

4 SAPs in IC

4.1 Introduction to SAP and Its Characteristics

SAPs are advanced materials that play a crucial role in modern concrete technology, offering unique properties and functionalities to improve the performance and durability of concrete structures. The properties of typical SAPs are presented in Table 2. SAPs are synthetic polymers with the ability to absorb and retain large amounts of water relative to their own weight, making them highly effective as IC agents in concrete. One of the key characteristics of SAP is its exceptional water absorption capacity, which allows it to absorb several hundred times its weight in water. This characteristic allows SAP to serve as a moisture reservoir within the concrete matrix, supporting ongoing hydration of the cementitious components and facilitating ideal curing conditions [4]. By regulating internal moisture levels, SAPs can control shrinkage, minimize cracking, and improve strength and longevity of concrete structures.

4.2 Mechanism of Water Absorption and Release by SAP in Concrete

SAPs are generally made up of hydrophilic polymers that are long-chained and have functional groups such as carboxylates or sulfonates. These groups have a strong attraction to water molecules because they form hydrogen bonds. While mixing, the water-attracting hydrophilic groups on SAPs easily draw in water molecules from the surrounding mixing water. This results in a gradient of concentration, with more water molecules concentrated near SAP and less within the polymer network. This

Table 2 Properties of SAPs

S. no	Property	Description	Typical values/ranges
1	Absorption capacity	The ability to absorb and retain moisture relative to its own weight, crucial for IC	100–1000 times its weight in water
2	Particle size	The size of SAP particles, which affects their distribution and performance in concrete	100–1000 μm
3	Swelling ratio	The ratio of the volume of SAP in the swollen state to its volume in the dry state	10–50 times
4	Release mechanism	The ability of SAP to release absorbed water over time, aiding the curing process	Gradual, controlled release
5	Compatibility with concrete mix	The ability of SAP to blend with cement, aggregates, and admixtures without adverse effects	High compatibility with standard concrete components
6	Chemical stability	The stability of SAP under the alkaline conditions of concrete	Stable in pH range 10–12
7	Biodegradability	The degree to which SAP breaks down under environmental conditions	Varies (often low biodegradability)
8	Cost	The relative cost of using SAP in concrete mixes	Moderate to high
9	Impact on workability	The effect of SAP on the fresh properties of concrete, such as slump and flow	May reduce workability (requires adjustments in mix design)
10	Impact on mechanical properties	The effect of SAP on hardened concrete properties like compressive strength and durability	Can improve durability; minor impact on strength

gradient causes water molecules to move into the SAP structure through diffusion. When water molecules enter the SAP network, they get intertwined with the polymer chains. Moreover, water molecules can surround and dissolve the ionic groups found in SAPs, enhancing their ability to absorb. As the outer moisture from concrete dries up, the volume of water molecules surrounding SAPs reduces. This creates a new concentration gradient, with a higher concentration of water inside SAPs compared to the surrounding concrete matrix [14]. The chemical interactions between the water molecules and the functional groups on SAPs require a certain amount of energy to break. This energy barrier prevents the uncontrolled release of all the absorbed water at once. Due to the existing concentration gradient, the water molecules inside SAPs begin to diffuse back out into the concrete matrix. This replenishes the moisture content within concrete, promoting continued hydration. SAPs are available in various forms, including powder, granules, and gel particles, and can be customized to meet specific project requirements. They are compatible with different types of cement, aggregates, and admixtures, allowing for flexible integration into concrete mixtures without compromising workability or setting characteristics.

4.3 Methods of Incorporating SAP into Concrete Mixtures

Direct method involves adding dry SAP powder directly to the other dry ingredients of the concrete mix during the initial batching process. For optimal distribution, it is recommended to add SAPs after the dry cementitious materials (cement and SCMs) have been thoroughly blended. This ensures SAP particles become well-coated with the cement particles, promoting better dispersion throughout the mix. One potential challenge with the dry method is the possibility of clumping, particularly if SAP powder is not well-dispersed. Insufficient mixing can lead to uneven distribution of SAPs within concrete, potentially compromising their effectiveness.

Pre-wetted method involves pre-wetting SAPs with a measured amount of water before incorporating them into the concrete mix [15]. SAPs are typically soaked in water for a specific duration, allowing them to absorb a significant amount of water and swell. The exact soaking time depends on the specific type of SAP used. The pre-wetted SAPs are then added to the concrete mix during the mixing process. This method can lead to a more uniform distribution of SAPs throughout concrete compared to the dry method.

4.4 Influence of SAP on Fresh and Hardened Concrete Properties

The effect of SAPs on workability can be seen in two ways; at lower dosages, SAPs might have minimal impact or even slightly improved workability due to their initial lubrication effect. However, at higher dosages, they can absorb some of the mixing water, leading to a stiffer and less workable mix [16]. Careful dosage optimization is crucial to maintain good workability. Similar to workability, the impact of SAPs on setting time can vary depending on the dosage. The presence of SAP in concrete mixtures may slightly extend the initial and final setting times, owing to the absorption of water by the polymer particles. However, this effect is generally minimal and can be managed through appropriate dosage adjustment. SAP can mitigate bleeding and segregation in fresh concrete by absorbing excess mixing water and enhancing cohesion. This results in improved homogeneity and reduces the risk of surface defects in the hardened concrete.

SAP can positively influence the strength development of hardened concrete by promoting optimal hydration conditions [17]. The continuous release of absorbed water ensures complete hydration of cementitious materials, leading to improved compressive and flexural strengths over time. Shrinkage cracking is a major concern in concrete construction. SAPs can noticeably reduce the risk of shrinkage cracking by providing a readily available internal water source throughout the critical early hydration stages [18]. This can lead to a more crack-resistant and durable concrete structure. SAP contributes to the durability of hardened concrete by improving its resistance to various deteriorative mechanisms, including freeze–thaw cycles, chemical attack, and ASR. The controlled release of moisture from SAP promotes the formation of dense, less permeable concrete with increased resistance to penetration by harmful substances, resulting in longer service life and reduced maintenance requirements. The presence of SAP in hardened concrete results in a reduction in porosity and permeability, which is beneficial for enhancing durability. By filling voids and capillary pores within the concrete matrix, SAP can reduce the ingress of water, chloride ions, and other deleterious substances, thereby improving the resistance of concrete to deterioration in aggressive environments.

4.5 Practical Applications of SAP

SAPs also provide similar benefits as of PEGs. Concrete placed in thin sections or applied through spraying techniques can dry out rapidly. During hot weather, concrete loses moisture rapidly due to evaporation. SAPs can act as a buffer, releasing internal moisture to compensate for external drying and promoting optimal hydration [19]. By providing an internal water source, SAPs help minimize shrinkage cracks that can compromise the structural integrity and durability of concrete.

4.6 Challenges Associated with SAP in ICC

Dosage Optimization: Obtaining the optimal dosage for a specific concrete mix is crucial. Insufficient quantities could lead to insufficient hydration, whereas an oversupply can affect how easily the material can be worked with or the time it takes to set.

Cost-effectiveness: While offering advantages, the cost of SAPs needs to be factored into the overall project budget.

Long-term Performance: Although research on the long-term performance of SAPs in concrete is ongoing, further investigations are necessary to fully understand their long-term effects on the concrete matrix.

5 LWAs for IC

5.1 Introduction to LWAs and Their Properties

LWAs are a key component in the production of lightweight concrete, offering unique properties and benefits that enhance the performance and durability of concrete structures. Unlike traditional aggregates such as gravel and sand, LWAs are characterized by their low density and high porosity, resulting in concrete with reduced weight and improved thermal and acoustic insulation properties. One of the

Table 3 Properties of LWAs

S. no	Property	Description	Typical values or ranges
1	Type of aggregate	Various types of LWAs used for IC	Expanded shale, clay, slate, pumice, perlite
2	Density	The mass per unit volume of LWAs, which is lower than normal-weight aggregates (NWAs)	300–1200 kg/m^3
3	Water absorption capacity	The ability of LWAs to absorb and retain water, contributing to IC	5–25% by weight (depending on the type of LWAs)
4	Porosity	The volume of pores within LWAs, which affects its water absorption capacity	30–60%
5	Strength	The compressive strength of LWAs, which is lower compared to NWAs	5–25 MPa
6	Compatibility with concrete mix	The ability of LWAs to blend with cement and other aggregates without causing adverse effects	High compatibility with standard concrete components
7	Thermal conductivity	The ability of LWAs to conduct heat, which is lower compared to NWAs	0.1–0.7 W/m K
8	Specific gravity	The ratio of density of LWAs to density of water, which is lower than that of NWAs	0.5–2.0

primary advantages of LWAs is their ability to internalize moisture and gradually release it during the hydration process of concrete, a property known as IC [5]. This IC mechanism can mitigate early-age shrinkage and cracking in concrete by ensuring continuous hydration of cementitious materials, even in conditions where external curing may be inadequate or impractical. Table 3 lists the typical properties of LWAs.

5.2 *Water Release from Internal Pores of LWAs to Support Hydration*

The key to understanding the water release mechanism lies in the inherent structure of LWAs. Unlike traditional NWAs that are dense and solid, LWAs are typically characterized by a highly porous internal structure. These pores can be large and interconnected and/or smaller and isolated. In some LWAs like expanded clay or shale, the pores are relatively large and interconnected, resembling a network of tiny channels. Other LWAs, such as some manufactured aggregates, might have smaller and more isolated pores within their structure.

During concrete mixing, water in the mix readily interacts with LWA particles. Due to the capillary action within the pores and the inherent attraction between water molecules and LWA surface, aggregates absorb a substantial amount of this mixing water. As concrete cures and dries, external moisture evaporates from the concrete surface. This creates a crucial difference in water concentration—a higher concentration of water molecules exists within LWA pores compared to the surrounding concrete matrix. This difference in concentration drives the release of water from LWAs. The release mechanism primarily relies on diffusion. Water molecules within the larger pores of LWAs can readily diffuse outward due to the concentration gradient. Additionally, in some cases, pressure gradients can also play a role. As the surrounding concrete dries and shrinks slightly, it can exert a slight pressure on the trapped water within LWAs, further aiding in its release.

5.3 Incorporation Techniques and Considerations for Using LWAs in Concrete

Direct batching is the simplest method, which involves the addition of dry LWAs directly to the other dry ingredients of the concrete mix during the initial batching process. For optimal distribution, it is recommended to add LWAs after the dry cementitious materials (cement and SCMs) have been thoroughly blended. This helps ensure LWA particles become well-coated with the cement particles, promoting better dispersion throughout the mix [20]. In some cases, pre-wetting LWAs with a measured amount of water before incorporation can be beneficial. Pre-wetting can potentially improve the initial dispersion of LWAs within the mix, particularly for aggregates with high absorption capacity. Additionally, it can reduce the overall water demand of the mix, as some of the mixing water has already been absorbed by LWAs. Pre-wetting introduces an additional step to the mixing process and requires careful control of the pre-wetting water content to avoid impacting the water-to-cement (w/c) ratio.

5.4 Effects of LWAs on Fresh and Hardened Concrete Properties

LWAs play a significant role in shaping the behavior of ICC, impacting both its fresh and hardened states. LWAs can influence the workability of fresh concrete. They tend to absorb water from the mixture, which can affect the overall water content and consistency of the concrete mix. Adjustments may be necessary to maintain the desired workability levels. On the other hand, incorporation of LWAs can also enhance workability due to their smoother, more spherical shape compared to traditional NWAs. However, maintaining the desired workability requires careful mix design considerations, including proper LWA gradation and content optimization. Since LWAs are pre-saturated for IC, they contribute less mixing water to the concrete mix compared to NWAs. This might necessitate slight adjustments to the w/c ratio to achieve the target slump [21]. LWA concrete demonstrates a lower density compared to normal-weight concrete, owing to the inherent lightness of aggregates. This reduction can be substantial depending on the replacement level of LWAs.

The incorporation of LWAs can affect the strength development of concrete and may vary based on the materials and mix type. While some studies reported minimal influence, others observed a slight decrease in strength with higher LWA replacement ratios. Employing SCMs or refining the mix design can mitigate this effect. LWAs typically have lower strength compared to traditional aggregates; the overall strength of concrete can still meet the design requirements when properly proportioned and cured. The pre-saturated LWAs function as internal reservoirs, gradually releasing water to cement paste as hydration progresses. This internal water source reduces water loss from the concrete matrix, significantly decreasing drying shrinkage and the risk of early-age cracking. The improved hydration facilitated by IC with LWAs can lead to a denser microstructure [22]. This denser structure reduces permeability, enhancing concrete durability against aggressive environments.

5.5 Real-World Applications of LWAs for IC

In transportation infrastructure, bridge decks and roadways are subjected to significant stress due to traffic loads and environmental factors. Incorporating LWAs in concrete for IC can enhance the durability of these structures by reducing shrinkage and cracking. This, in turn, extends the service life of the infrastructure and minimizes maintenance needs. Repair and rehabilitation of existing concrete structures often involve challenges such as limited access, tight schedules, and compatibility with existing materials. IC using LWAs can facilitate these projects by improving the bond between new and existing concretes, reducing shrinkage-induced stresses, and accelerating the overall repair process. By reducing the consumption of traditional natural resources and energy-intensive materials, such as cement, LWAs contribute to lower embodied carbon and overall environmental impact of construction projects.

5.6 Challenges of Utilizing LWAs in Concrete Construction

Striking a balance between the benefits of reduced shrinkage and maintaining desired strength and workability is crucial. This can be achieved by optimizing LWA content and the overall mix design. The lower density of LWA concrete might necessitate adjustments in design procedures compared to normal-weight concrete. While LWAs offer great advantages in concrete construction, particularly for ICC, their application presents certain challenges that require attention. The smoother, lower-density nature of LWAs can initially improve workability. However, segregation becomes a concern, where LWAs separate from cement paste, compromising homogeneity. Higher LWA content can lead to a slight reduction in compressive strength compared to normal-weight concrete. This can be mitigated by incorporating SCMs like silica fume or fly ash or by refining the mix design to achieve a denser matrix. While ICC with LWAs generally reduces drying shrinkage, challenges can arise. Improper mix design or highly absorbent LWAs might lead to insufficient IC, still resulting in shrinkage cracks. Thus, careful selection of LWAs with proper absorption capacity and optimization of the mix for optimal internal water release are crucial.

6 Challenges and Future Directions

ICC offers a promising approach for achieving high-performance, durable concrete with reduced shrinkage cracking. However, there are still challenges to overcome and promising directions to explore in the future.

6.1 Challenges

Mix Design Optimization: Optimizing the mix design for ICC can be complex. Factors like the type and amount of IC agent, its compatibility with other mix components, and the impact on fresh and hardened concrete properties require careful consideration. One of the primary challenges is the selection and compatibility of IC agents with different concrete mixtures. While materials like LWA, PEG, and SAP offer IC capabilities, their interaction with various cementitious systems and admixtures requires careful consideration. Future research should focus on identifying optimal combinations of IC agents that enhance performance without compromising the integrity of concrete.

6.1.1 Long-Term Performance

Assessing the long-term performance and durability of ICC remains a key area of research. While IC agents can mitigate early-age issues such as shrinkage and cracking, their effectiveness over the lifespan of concrete structures needs to be thoroughly evaluated. Future directions should include comprehensive field studies, monitoring programs, and predictive modeling to assess the durability performance of ICC under real-world conditions and inform design practices accordingly.

6.1.2 Quality Control

The effectiveness of ICC heavily relies on the quality and consistency of the IC agent, particularly for agents like SAP or LWA with varying properties. Strict quality control measures are essential to ensure consistent performance.

6.1.3 Cost-Effectiveness

Depending on the chosen IC agent, ICC can sometimes be more expensive than conventional concrete. Striking a balance between the cost and benefits achieved is crucial.

6.1.4 Standardization and Acceptance

Wider industry acceptance and standardized guidelines for ICC implementation are needed to encourage its broader adoption in construction projects.

6.2 Future Directions

6.2.1 Advanced IC Agents

Research on developing new and improved IC agents with targeted properties, such as controlled water release rates or compatibility with a wider range of mix designs, is ongoing.

6.2.2 Modeling and Simulation

Utilizing advanced modeling and simulation techniques can help optimize ICC mix designs and predict the performance of concrete with different IC agents.

6.2.3 Life-Cycle Cost Analysis

Performing life-cycle cost analyses that consider not just the initial cost of ICC but also the benefits in terms of reduced maintenance, improved durability, and potentially lower dead loads due to LWAs can strengthen the case for wider adoption.

6.2.4 Sustainability Focus

As the construction industry strives for more sustainable practices, ICC's potential to reduce water usage through improved hydration efficiency and potentially lower cement content due to optimized mix designs aligns well with these goals.

By addressing the existing challenges and actively exploring these promising future directions, ICC can become a mainstream construction material, leading to the development of more durable, crack-resistant, and sustainable concrete structures.

7 Conclusions

This chapter has delved into the innovative concept of ICC, exploring its fundamental principles, benefits, and the various agents used to achieve IC. By integrating these agents into the concrete mix, ICC addresses many of the shortcomings associated with traditional curing methods. The following conclusions can be drawn from this study:

(a) ICC has emerged as a transformative approach to enhance concrete performance by promoting improved hydration, reducing drying shrinkage, and ultimately leading to more durable and crack-resistant structures. This chapter explored the utilization of three prominent IC agents: PEG, SAP, and LWA.
(b) Each agent offers distinct advantages. PEG provides controlled water release, but its influence on setting time necessitates careful consideration. SAP boasts exceptional water absorption capacity, but cost and compatibility with other mix components require evaluation.
(c) Evaluating ICC performance includes analyzing the properties of both fresh and hardened concretes. Key metrics include workability, water requirement, strength, drying shrinkage, and permeability. Choosing the best IC agent depends on the project's needs, taking into account factors such as desired properties of fresh concrete, target strength, shrinkage reduction requirements, durability considerations, and cost efficiency.
(d) Several challenges persist, such as optimizing mix design, maintaining quality control, and achieving cost-efficiency for broader use. However, promising future prospects present solutions to these issues. Ongoing research into advanced internal curing agents with specific properties, along with the use of modeling and simulation tools for optimized mix designs, shows great potential. Furthermore, life-cycle cost assessments that factor in long-term advantages and the alignment of ICC with sustainable construction practices are paving the way for its wider acceptance.
(e) In summary, ICC, through its range of internal curing agents, offers a compelling approach to producing high-performance concrete. By overcoming current obstacles and actively pursuing future innovations, ICC has the potential to

transform the construction industry, leading to the creation of more durable, sustainable, and crack-resistant concrete structures that endure over time.

References

1. ACI Committee, ACI 308-213 R report on internally cured concrete using Prewetted absorptive lightweight aggregate. Am. Concr. Inst. **28**, 0–15 (2015)
2. R. Henkensiefken, J. Castro, D. Bentz, T. Nantung, J. Weiss, Water absorption in internally cured mortar made with water-filled lightweight aggregate. Cem. Concr. Res. **39**, 883–892 (2009). https://doi.org/10.1016/j.cemconres.2009.06.009
3. V. Kastro Kiran, D. Sathyan, Unveiling the combined influence of higher molecular weight polyethylene glycol and superplasticizer chemistry on fresh, mechanical, and microstructural performance of internally cured mortar. Mater. Res. Express **11** (2024). https://doi.org/10.1088/2053-1591/ad5c30
4. Z.L. Zhang, X.M. Kong, Effect of super-absorbent polymer on pore structure of hardened cement paste in high-strength concrete. J. Chin. Ceram. Soc. **41**, 1474–1480 (2013)
5. S. Yang, L. Wang, Effect of internal curing on characteristics of self-compacting concrete by using fine and coarse lightweight aggregates. J. Mater. Civ. Eng. **29**, 1–11 (2017). https://doi.org/10.1061/(asce)mt.1943-5533.0002044
6. J.J. Beaudoin, H. Dramé, L. Raki, R. Alizadeh, Formation and properties of C-S-H-PEG nanostructures. Mater. Struct. **42**, 1003–1014 (2009). https://doi.org/10.1617/s11527-008-9439-x
7. T. Udayabanu, N.P. Rajamane, C. Makendran, R. Gobinath, S. Chandra Chary, Self-curing concrete using water-soluble polymer for developing countries. IOP Conf. Ser. Mater. Sci. Eng. **981**, 032088 (2020). https://doi.org/10.1088/1757-899X/981/3/032088
8. S. Salimi, M. Nasr-Esfahani, S.A. Umoren, E. Saebnoori, Complexes of imidazole with poly(ethylene glycol) as a corrosion inhibitor for carbon steel in Sulphuric acid. J. Mater. Eng. Perform. **24**, 4696–4709 (2015). https://doi.org/10.1007/s11665-015-1788-3
9. V. Kastro Kiran, D. Sathyan, K.B. Anand, P.R. Devi, Strength and durability properties (effect of salts) of internal curing concrete. IOP Conf. Ser. Mater. Sci. Eng. **872**, 012115 (2020). https://doi.org/10.1088/1757-899X/872/1/012115
10. N. Hamzah, H.M. Saman, M.H. Baghban, A.R.M. Sam, I. Faridmehr, M.N.M. Sidek, O. Benjeddou, G.F. Huseien, A review on the use of self-curing agents and its mechanism in high-performance cementitious materials. Buildings **12**, 1–26 (2022). https://doi.org/10.3390/buildings12020152
11. D. Sarbapalli, Y. Dhabalia, K. Sarkar, B. Bhattacharjee, Application of SAP and PEG as curing agents for ordinary cement-based systems: Impact on the early age properties of paste and mortar with water-to-cement ratio of 0.4 and above. Eur. J. Environ. Civ. Eng. **21**, 1237–1252 (2017). https://doi.org/10.1080/19648189.2016.1160843
12. R.K. Dhir, P.C. Hewlett, T.D. Dyer, Durability of "self-cure" concrete. Cem. Concr. Res. **25**, 1153–1158 (1995). https://doi.org/10.1016/0008-8846(95)00107-N
13. M. Amin, A.M. Zeyad, B.A. Tayeh, I. Saad Agwa, Engineering properties of self-cured normal and high strength concrete produced using polyethylene glycol and porous ceramic waste as coarse aggregate. Constr. Build. Mater. **299**, 124243 (2021). https://doi.org/10.1016/j.conbuildmat.2021.124243
14. L. Montanari, P. Suraneni, M.T. Chang, C. Villani, J. Weiss, Absorption and desorption of superabsorbent polymers for use in internally cured concrete. Adv. Civ. Eng. Mater. **7**, 547 (2018). https://doi.org/10.1520/ACEM20180008
15. K. Suganya, V. Poornima, Study on absorbency of superabsorbent polymer in water and cement pore fluid and its incorporation in cement mortar. Int. J. Civ. Eng. Technol. **8**, 1131–1140 (2017)
16. D.P. Bentz, P. Lura, J.W. Roberts, Mixture proportioning for internal curing. Concr. Int. **27**, 35–40 (2005). http://concrete.nist.gov/~bentz/Mixpropfin/CI2702Bentz.pdf

17. H. Beushausen, M. Gillmer, M. Alexander, The influence of superabsorbent polymers on strength and durability properties of blended cement mortars. Cem. Concr. Compos. **52**, 73–80 (2014). https://doi.org/10.1016/j.cemconcomp.2014.03.008
18. A.M. Soliman, M.L. Nehdi, Effect of drying conditions on autogenous shrinkage in ultra-high performance concrete at early-age. Mater. Struct. Constr. **44**, 879–899 (2011). https://doi.org/10.1617/s11527-010-9670-0
19. V. Poornima, R. Venkatasubramani, V. Sreevidya, V. Vignesh, K. Adithan, Study on SAP with different absorption capacities and its effects on self curing HPC. IOP Conf. Ser. Mater. Sci. Eng. **872**, 012174 (2020). https://doi.org/10.1088/1757-899X/872/1/012174
20. D. Shen, M. Wang, Y. Chen, T. Wang, J. Zhang, Prediction model for relative humidity of early-age internally cured concrete with pre-wetted lightweight aggregates. Construct. Build. Mater. **144**, 717–727 (2017). https://doi.org/10.1016/j.conbuildmat.2017.03.088
21. N. Yadav, S.V. Deo, G.D. Ramtekkar, Mechanism and benefits of internal curing of concrete using light weight aggregates and its future prospects in Indian construction industry. Int. J. Civ. Eng. Technol. **8**, 323–334 (2017)
22. Y.W. Jeong, T.H. Koh, K.S. Youm, J. Moon, Experimental evaluation of thermal performance and durability of thermally-enhanced concretes. Appl. Sci. **7**, 811 (2017). https://doi.org/10.3390/app7080811

Experimental Analysis of Spent Wash as a Waste Material in Mortar

Bhagyashree Khartode, Manisha Shewale, and Nitin Shinde

1 Introduction

India's rapid economic expansion necessitates high-quality roadways to foster continued growth. The selection of pavement materials plays a crucial role in determining road and infrastructure construction costs. Road surfaces typically employ either bitumen or concrete. Concrete pavements are formulated using various cement grades, aggregates, sand, admixtures, and water. In contrast, bitumen pavements incorporate bitumen or tar, aggregates, and fillers such as basalt dust or lime. Pavements are critical infrastructure components that facilitate transportation across various platforms, including roads, runways, parking areas, and driveways. These engineered structures are fundamental to everyday life, supporting commerce, trade, and national defense efforts. Surface transportation dominates globally, with a nation's paved road network often indicating its level of development. The road construction sector remains vital in developing countries and is anticipated to grow in significance as infrastructure progresses. Additionally, this industry is expected to gain prominence in developed nations as their infrastructure evolves. The construction sector noticeably enhances societal wellbeing by improving individual quality of life [1]. Globally, substantial quantities of waste are generated, posing considerable risks due to their potentially toxic, caustic, combustible, chemically reactive, easily ignited, and pathogenic properties. These materials are typically disposed of in landfills, resulting in environmental degradation and potential health hazards. Consequently, incorporating waste materials into concrete represents a more responsible and beneficial approach [2]. Various industries, including mining, manufacturing, agriculture, power generation, iron and steel production, and electronics manufacturing, produce significant volumes of solid waste. These waste streams

B. Khartode (✉) · M. Shewale · N. Shinde
Department of Civil Engineering, Dr. D. Y. Patil Institute of Technology, Pimpri, Pune, India
e-mail: bhagyashree.khartode@dypvp.edu.in

A. Bahrami (ed.), *Advances in Sustainable Concrete for Construction*,
https://doi.org/10.1007/978-3-031-85052-3_8

contain numerous hazardous components known for their flammability, chemical reactivity, combustibility, corrosiveness, and infectious nature. The improper disposal and management of these wastes have resulted in substantial economic consequences [3, 4].

Large amounts of liquid and solid wastes are produced by water-intensive industries like distilleries and sugar processing. Biomethane spent wash (BMSW) is a notable semisolid residue that is frequently encountered in sugar and distillery operations among these byproducts. After sugar is extracted from sugarcane juice using a multistep evaporation process, a thick, dark substance called molasses is left over. In neighboring distilleries, this molasses is then fermented to create bioethanol. However, a large amount of water is used in this process. Around 12–14 L of water are needed for every liter of ethanol, and a considerable amount of trash is produced in the form of spent wash [5]. BMSW wastewater is now treated by utilizing a variety of methods with the goal of reducing its coloration and chemical oxygen demand (COD) [6]. Ongoing research is still being done on the environmental effects of disposing of treated BMSW in water bodies. This wastewater stands out owing to its high levels of total organic carbon (TOC), biochemical oxygen demand (BOD), and COD [7]. If spent wash, a byproduct of the distillation process used in the sugar industry, is not properly handled before being released into water bodies, it can be hazardous [8]. Spent wash is a dark effluent that is created during the distillation process of rectified spirits in molasses-based industries. Decomposable organic materials and dissolved salts are present in significant concentrations. Maintaining environmental beauty requires the management of industrial waste to reduce its negative effects on the environment. The complicated sludge created during the electrocoagulation treatment of distillery wasted wash calls for suitable disposal techniques. One strategy is to produce nonstructural building blocks by utilizing the sludge produced by electrocoagulation as a partial cement substitute [9]. Waste from distilleries can be used in concrete as an additive. Concrete specimens intended for M20 grade were mixed with different percentages of distillery spent wash (DSW) for experimental reasons, ranging from 0 to 1.25%. Slump cone and compaction factor tests were employed to evaluate the workability of concrete containing distillery waste wash, while compressive and flexural tests were used to examine the strength properties [10]. The expansion of industrial activities has resulted in a persistent rise in environmental contamination, posing a noticeable threat to soil, water, and air quality. The extensive effects of industrialization have led to the discharge of substantial volumes of wastewater, exacerbating environmental concerns. This industrial effluent, originating from various sources, is increasingly utilized for agricultural irrigation. However, effective wastewater management remains a considerable challenge in mitigating potential adverse impacts on soil and environmental ecosystems. A notable example of industrial wastewater is DSW from the sugarcane industry [11]. Numerous independent studies have explored the elimination of pollutants and heavy metals by employing various adsorption methods and adsorbent materials, with fly ash and activated carbon emerging as the most economically viable options. Analysis suggests that a 50% concentration of DSW appears optimal for further investigation [12]. Concrete pavements primarily experience flexural stresses, leading to a greater emphasis on flexural strength specifications rather than compressive strength in the design of concrete mixes for pavement construction [13]. Agricultural byproducts offer a viable alternative for cement substitution in cementitious composites, contributing to the reduction of CO_2 emissions associated with cement production. Specifically, processed waste tea ash can be employed as a partial cement replacement in paving block formulations at varying dosage levels [14]. Studies demonstrate that while high doses (>250 m^3/ha) of spent wash negatively affect crop growth and soil fertility, lower doses (125 m^3/ha) significantly enhance germination, growth, and yield of dryland crops. Moreover, combining spent wash with organic amendments has proven effective in dryland conditions. It has also been observed that substantial quantities of soluble salts are leached from calcareous and high pH sodic soils treated with spent wash [15]. Concrete workability is essential for efficient pouring. Construction work becomes extremely challenging when concrete lacks workability. The addition of waste foundry sand reduces workability, resulting in decreased fluidity and slump value of fresh concrete [16].

1.1 Spent Wash

Across various industries, the manufacturing sector produces substantial amounts of waste. The main issue is finding ways to break down this waste without harming the environment. In the molasses-based industry, a brown effluent called spent wash is produced during the distillation of rectified spirits. Spent wash offers a potential use in concrete production [17] due to its high levels of decomposable organic matter and dissolved salts. Using DSW extensively in construction helps conserve cement, improves concrete mixture quality, and tackles environmental concerns. It is also utilized to enhance soil properties in road construction. When industries discharge spent wash into water bodies as waste, it leads to water pollution and raises the water pH, which is detrimental to aquatic life. Some industries decompose spent wash in burners to create ash. This process releases large quantities of air pollutants, which are highly acidic and can cause acid rain, harming the environment. Table 1 lists the characteristics of fresh spent wash (FSW), while Table 2 shows the chemical properties of spent wash after being stored for 6 months.

Table 1 Physiochemical properties of FSW

Sr. no	Properties	Test result
1	Color of FSW	Dark brown
2	Specific gravity	1.2
3	pH	6.7
4	Organic carbon %	3.7
5	Dry matter	18–26%
6	Pigments	4–8%
7	Odor	Unpleasant burnt sugar

Table 2 Chemical properties of spent wash after 6 months

Sr. no	Properties	Test result	Standard value range	Standard code
1	pH	8	6.5 to 8.5	IS 3025 Part 11
2	COD in Mg/lit	340,000	250 (Max)	APHA
3	BOD at 27 °C for 3 days in Mg/lit	163,000	50 (Max)	APHA

Fig. 1 Spent wash in liquid and powder forms

Several key characteristics of FSW are observed in Table 1. Spent wash is initially characterized by a dark brown color, a specific gravity of 1.2, and a slightly acidic pH of 6.7. Additionally, it contains 3.7% organic carbon and exhibits a dry matter content ranging from 18% to 26%. The presence of pigments contributes to their coloration, while an unpleasant burnt sugar odor is detected.

In contrast, Table 2 reveals the chemical properties of spent wash after 6 months of storage. The most notable change is the increase in pH from 6.7 to 8.00, illustrating a shift toward alkalinity. This change in pH may be attributed to the microbial degradation of organic compounds present in the spent wash over time. Moreover, the COD and BOD values display a substantial increase, with COD reaching 340,000 Mg/lit and BOD at 270 °C for 3 days reaching 163,000 Mg/lit. These values exceed the maximum permissible limits of 250 Mg/lit for COD and 50 Mg/lit for BOD, as per the standard guidelines.

Overall, the transformation of spent wash from its initial state to its condition after 6 months reflects the microbial decomposition of organic matter, leading to pH changes and the accumulation of oxygen-demanding substances (Fig. 1).

2 Materials

Materials play an important role for obtaining strength in any concrete or mortar of different proportions. The following materials are used to prepare molds of concrete and mortar with different proportions (Fig. 2).

Fig. 2 Materials used for preparing samples. (**a**) Cement. (**b**) Silica sand. (**c**) Spent wash. (**d**) Silica grit. (**e**) Steel fiber. (**f**) Basalt fiber. (**g**) Plasticizer. (**h**) Latex

Table 3 Properties of OPC

Chemical composition	Test result (%)	Physical properties	Test result
CaO	58.73	Specific gravity	3.15
SiO_2	21.45	Blaine specific surface area	3145 cm^2/g
Al_2O_3	9.27	Initial setting time	95 min
MgO	2.91	Final setting time	190 min
Fe_2O_3	3.87	28-day compressive strength	55.37 MPa

2.1 Cement

Cement, a crucial chemical substance in construction, functions as a binding agent that sets, hardens, and adheres to various materials, effectively joining them together. It is seldom used on its own; rather, it is typically mixed with aggregate materials such as sand and gravel. When combined with fine aggregate, cement creates mortar for masonry applications, while mixing it with sand and gravel produces concrete. In the preparation of test specimens, ordinary Portland cement (OPC) grade 53, which complies with IS: 269-2015 standard [18], was employed. The characteristics of OPC are presented in Table 3.

2.2 Silica Sand

Composed primarily of quartz, silica sand is a granular substance utilized as fine aggregate, containing minor amounts of coal and clay and sometimes other minerals. This material, also referred to as quartz sand, white sand, or industrial sand, is

Table 4 Properties of fine aggregates

Properties	River sand	Silica sand
Specific gravity	2.65	2.53
Fineness modulus	2.77	2.69
Grading zone	Zone II	Zone II
Bulk density	1630 kg/m^3	1580 kg/m^3

predominantly silicon dioxide (SiO_2). The presence of silica sand can initiate crevice corrosion on metal surfaces. Being one of the most abundant minerals in the earth's crust, silica sand finds extensive application across various industries owing to its beneficial properties. For this study, silica grit and silica sand were sourced from local mines supplying a silica cement production facility. The characteristics of both river sand and silica sand are summarized in Table 4.

2.3 Fibers

Steel fiber serves as an effective reinforcement material, enhancing concrete toughness, durability, and flexural strength. This results in improved concrete performance during construction. Worldwide, steel fiber is widely used to reinforce precast concrete products due to its inherent strengthening qualities. The process of manufacturing basalt fibers involves melting basalt-containing rocks and transforming the resulting liquid into fibrous material. Basalt, an igneous rock, serves as the primary raw material. The main energy requirement for preparing basalt raw materials for fiber production occurs naturally. Various types of basalt fibers are produced, including continuous, staple, and superthin varieties. Continuous basalt fibers are utilized in the creation of reinforcing materials, composite products, fabrics, and nonwoven materials. The physical properties of these fibers are exhibited in Table 5.

2.4 Plasticizer

Concrete additives known as plasticizers are used to reduce viscosity and enhance workability. For mortar applications, superplasticizers contribute to increased flexibility and improved consistency maintenance. This can potentially enable bricklayers to enhance their productivity by laying more bricks. As outlined in Table 6, new-generation superplasticizers based on modified polycarboxylic ether were employed as a chemical admixture to improve workability.

Table 5 Properties of fibers

Physical properties	Steel fiber	Basalt fiber
Diameter	0.75 mm	0.1–0.2 mm
Cut length	50–60 mm	10–12 mm
Tensile strength	1100–1200 MPa	2700–3000 MPa
Elastic modulus	205 GPa	75–87 GPa
Density	7.85 g/cm^3	2.67 g/cm^3

Table 6 Properties of superplasticizers

Properties	Test result
Relative density	1.13
pH value	>6.0
Color	Light brown liquid
Chloride content (%)	<0.2

Table 7 Properties of SBR

Property	Test value
Specific gravity	1.02
Density (kg/L)	1
7-day compressive strength (MPa)	32
7-day flexural strength (MPa)	15
7-day tensile strength (MPa)	3.8

2.5 SBR Latex

Styrene–butadiene rubber (SBR) latex was utilized as an adhesive agent to enhance bonding strength. This synthetic polymer dispersion finds widespread use in various industrial sectors, particularly in the production of coatings, adhesives, sealants, and latex-based goods. SBR latex is produced through the polymerization of styrene and butadiene monomers, resulting in a copolymer that combines the attributes of both components. Table 7 provides a list of the physical properties of latex.

3 Preparations and Testing of Samples

3.1 Sample Preparation

Samples were prepared with varying percentages of FSW and stored spent wash (SSW) that had been aged for 6 months. In total, 27 samples were tested to evaluate cylindrical compressive strength, splitting tensile strength, and flexural strength for different proportions of FSW and SSW. Moreover, 189 samples were cast to examine the mechanical properties of mortar with different percentages of FSW and SSW.

Cylindrical molds that measure 60 mm in diameter and 120 mm in height were utilized to cast samples for testing compressive and splitting tensile strengths of

Fig. 3 Cylindrical and prism molds and preparation of specimens

Fig. 4 Testing of mortar specimens

mortar developed in this study. The prism specimen dimensions, which were prepared in accordance with ASTM C348 [19], were 40 × 40 × 160 mm. These molds were used to evaluate the flexural strength of mortar. The molds and the process of preparing the mortar specimens are shown in Fig. 3. Spent wash was utilized in 2%, 4%, and 6% of cement for casting the samples (Fig. 4).

3.2 Compression Test

The standards for conducting compressive strength tests on mortar specimens are mentioned in IS 4031 (Part 6) [20] and ASTM C31/C31M 09 [21]. These guidelines specify the dimensions for cylindrical mortar specimens used in testing, requiring a diameter of 60 mm and a height of 120 mm. The height-to-diameter ratio of cylindrical specimens is a critical factor in material testing, as it can affect test outcomes. This ratio, which is calculated by dividing the specimen's height by its diameter, is generally recommended to be between 1.0 and 2.0, based on industry norms and best practices. ASTM C31/C31M 09 [21] further stipulates that for compressive strength and splitting tensile strength testing, the height of cylindrical specimens should be twice their diameter. Adhering to these dimensional requirements for cylindrical mortar specimens is essential for obtaining precise and dependable test results in compressive strength testing. The following equation is used to calculate cylindrical compressive strength:

$$F = \frac{P}{A}$$

where

F: Compressive strength of specimen (N/mm^2).
P: Maximum load taken by specimen (N).
A: Cross-sectional area (mm^2).

3.3 Splitting Tensile Test

Specific protocols must be adhered to when performing splitting tensile strength tests on cylindrical samples measuring 60 mm in diameter and 120 mm in height to ensure precise and dependable outcomes. Splitting tensile strength measures a material's capacity to withstand tensile stresses, which arise when force is applied perpendicular to the material's surface. For cylindrical specimens, this property refers to the material's ability to resist splitting under tensile force applied along the cylinder's length. The examination involves steadily applying a load to the cylindrical sample at a regulated pace. This loading rate is critical as it can influence test outcomes. As indicated by the provided sources, the load should be applied smoothly and continuously, avoiding any abrupt shocks or jerks. The loading rate is specified to fall within 0.7 to 1.4 MPa/min according to ASTM C496/C496M-11 [22]. However, IS 516:1999 [23] suggests a slightly different range of 1.2 to 2.4 MPa/min for the loading rate.

The splitting tensile strength is obtained utilizing the equation below:

$$T = \frac{2P}{\pi DL}$$

where

T: Splitting tensile strength (N/mm^2).
P: Maximum load taken by specimen (N).
D: Diameter of specimen (mm).
L: Length of specimen (mm).

3.4 Flexural Test

The testing procedure outlined in IS 516:1999 [23] and ASTM C348–08 [19] involves the preparation of prism specimens with dimensions of 40 mm × 40 mm × 160 mm. During the test, a continuous load is applied to the specimen until it fails, ensuring that no sudden or jarring impacts occur. The recommended loading rates are 400 kg/min for 150 mm specimens and 180 kg/min for 100 mm specimens, which are designed to maintain uniform and consistent load application throughout the test. Additionally, British standards specify that the stress increase rate during loading should be 0.06 +/− 0.04 N/mm^2 per second.

The following equation is employed to calculate flexural strength:

$$F_b = \frac{PL}{bd^2}$$

where

F_b: Flexural strength (N/mm^2).
P: Maximum load taken by specimen (N).
L: Supported length (mm).
b: Width of specimen (mm).
d: Failure point depth (mm).

4 Results and Discussion

4.1 Workability Test

Figure 5 displays the workability test results of mortar samples containing FSW. The results demonstrate that higher percentages of spent wash improved sample workability. This enhancement is attributed to the adsorption of hydrophilic surfactants onto solid cement particles, altering their electrokinetic potentials. As a result, van der Waals forces within the cement–aggregate–water system decreased, slowing the coagulation of newly formed structures during cement–water interaction. Moreover, these surfactants optimized the contact between CaO, SiO_2, and water molecules, ultimately reducing internal friction and enhancing fluidity of the concrete mixture.

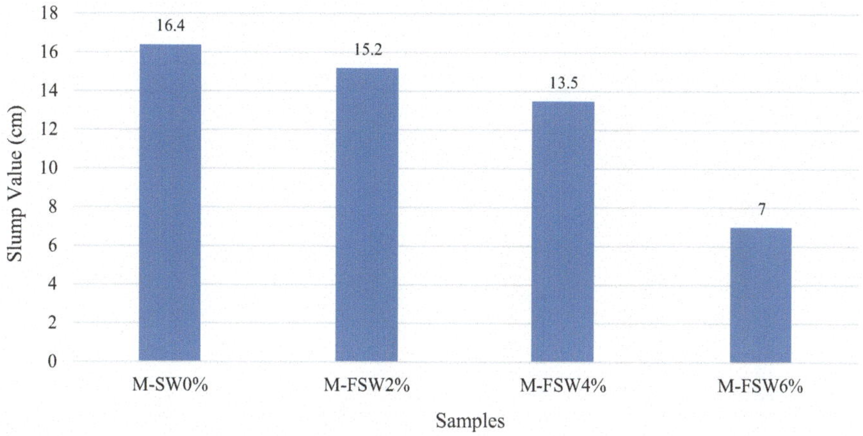

Fig. 5 Slump values of mortar samples

The concrete mix without FSW (M-SW0%) demonstrates the highest workability, with a slump value of 16.4 cm. As the FSW content increased, workability decreased: 2% FSW yielded a 15.2 cm slump, 4% FSW resulted in a 13.5 cm slump, and 6% FSW produced a 7 cm slump. The M-FSW6% sample, containing the highest FSW percentage, exhibited the lowest workability. This inverse relationship between the FSW content and workability, evidenced by declining slump values, is significant for the concrete mix design. It suggests that higher FSW concentrations lead to stiffer mixtures, potentially complicating the placement and finishing processes. Consequently, careful management of FSW proportions is essential to achieve an optimal balance between workability and other concrete properties.

4.2 Compressive Strength

Tests for compressive, splitting tensile, and flexural strengths were performed on mortar samples containing 0%, 2%, 4%, and 6% spent wash. Figure 6 depicts the compressive strength results of the samples at 3, 7, and 28 days.

The control sample (M-SW0%) illustrated a normal increase in compressive strength over time, reaching 41.11 MPa after 28 days. Incorporating 2% FSW substantially enhanced compressive strength across all curing periods, with a peak value of 56.3 MPa at 28 days, marking a 37% improvement over the control. However, further increasing the FSW content to 4% and 6% has resulted in a decrease in compressive strength, though values remained superior to those of the SSW samples.

Conversely, the addition of SSW considerably diminished the compressive strength for all the percentages and curing durations, with the highest recorded value being merely 1.2 MPa for the 2% SSW mixture at 28 days. This suggests that

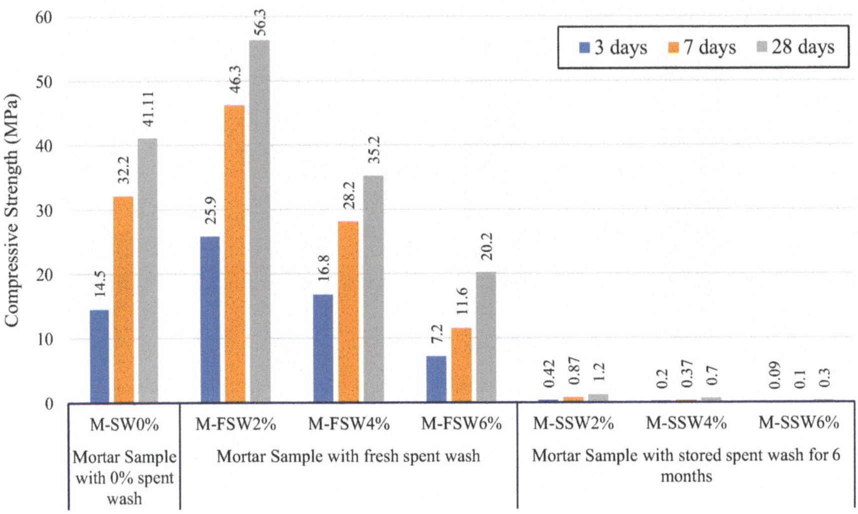

Fig. 6 Compressive strength results of mortar samples

SSW is less effective than FSW as a supplementary material in concrete. Notably, the minimal compressive strength values for the SSW mixtures indicate its unsuitability for structural applications requiring significant strength.

To summarize, FSW showed promise as a strength-enhancing additive at lower percentages, particularly at 2%. In contrast, SSW appeared to have a detrimental effect on compressive strength, rendering it unsuitable for such applications.

4.3 *Splitting Tensile Strength*

The splitting tensile strength results of the mortar samples are presented in Fig. 7. The control sample (M-SW0%) showed a characteristic increase in strength over time, reaching 5.5 MPa after 28 days. Adding 2% FSW substantially enhanced splitting tensile strength, achieving 6.1 MPa at 28 days, which represented a significant 10.9% improvement compared to the control. However, higher FSW percentages (4% and 6%) have resulted in a decrease in splitting tensile strength, with the 6% FSW mixture demonstrating a considerable reduction, only attaining 1.5 MPa at 28 days.

In comparison, SSW addition consistently decreased splitting tensile strength across all the levels. The highest value observed for the SSW mixtures was 0.174 MPa for the 4% SSW sample at 28 days, displaying that SSW did not effectively contribute to splitting tensile strength and is thus unsuitable for applications requiring high splitting tensile strength.

To summarize, the findings exhibited that 2% FSW addition improved splitting tensile strength of concrete, while higher percentages resulted in reduced strength.

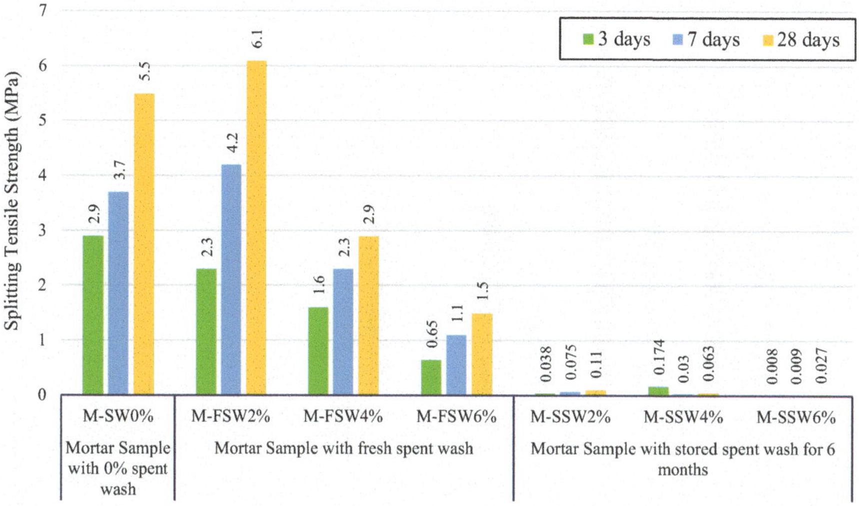

Fig. 7 Splitting tensile strength results of mortar samples

Conversely, SSW consistently weakened splitting tensile strength, indicating its ineffectiveness as a concrete additive for enhancing splitting tensile properties. These results underscored the potential of FSW at lower percentages for the strength improvement, while SSW is not appropriate for similar enhancements.

4.4 Flexural Strength

This study examined the flexural strength of various concrete mixtures after 3, 7, and 28 days of curing. These mixtures contained different proportions of spent wash, ranging from 0% to 6%. Figure 8 illustrates the flexural strength results of the mortar samples. The control sample (M-SW0%) demonstrated a typical increase in flexural strength over time, reaching 13.5 MPa at 28 days. Notably, the addition of 2% FSW significantly boosted flexural strength across all the curing periods, peaking at 15.3 MPa after 28 days, which represented an improvement of approximately 13.3% compared to the control sample.

However, increasing the FSW content to 4% and 6% has resulted in a decrease in flexural strength. The 4% FSW mixture achieved 9.16 MPa at 28 days, while the 6% FSW mixture reached 8.15 MPa, both falling below the control sample's 28-day flexural strength.

In contrast, the introduction of SSW at any percentage substantially reduced flexural strength. The highest value observed for the SSW mixtures was 0.77 MPa for the 2% SSW mixture at 28 days. This finding suggests that SSW is ineffective in enhancing mortar flexural strength, rendering it unsuitable for structural applications requiring higher flexural strength.

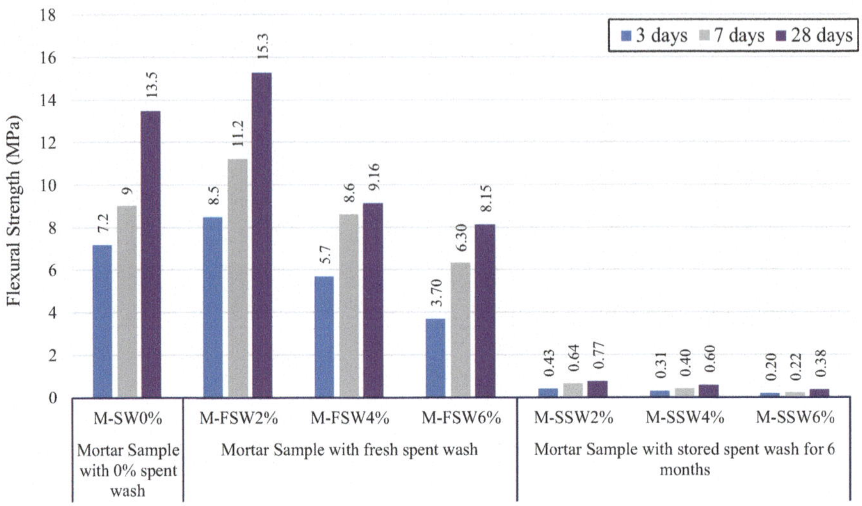

Fig. 8 Flexural strength results of mortar samples

To summarize, incorporating 2% FSW markedly enhanced concrete flexural strength, whereas higher percentages (4% and 6%) resulted in strength reduction. On the other hand, SSW consistently diminished flexural strength across all the tested percentages, underscoring its unsuitability as a concrete additive for improving flexural properties. These results highlight the potential of low-percentage FSW for boosting concrete flexural strength while showing the limitations of SSW for similar applications.

5 Conclusions

This chapter has investigated the mechanical properties of inorganic mortar containing various amounts of spent wash, utilizing different mixtures of sand, fibers, and cementitious materials. Spent wash, a sugar-based molasses, functions as a retarding agent in mortar. It serves as an effective air-entraining agent, enhancing mortar cohesion and workability, but negatively impacts compressive strength when used in high quantities. Experimental findings indicated that 2% is the ideal spent wash dosage, yielding favorable compressive, splitting tensile, and flexural strengths. While FSW appears to improve strengths initially, SSW exhibits a significant decrease in the strength properties over the curing duration. Key findings include the following:

- FSW initially enhanced the compressive strength of the mortar samples, but this effect lessened over time, as evidenced by decreased compressive strength at 28 days for higher FSW percentages. Conversely, SSW displayed a marked reduction in compressive strength compared to FSW, with substantially lower values across all the percentages. The highest strength of 56.3 kN/m^2 was observed with 2% FSW.

- At 28 days, the mortar samples without spent wash achieved 5.5 kN/m^2 splitting tensile strength, while those with 2% spent wash reached 6.1 kN/m^2. Strengths decreased in the samples with 4% and 6% spent wash compared to the 2% dosage. Spent wash stored for 6 months yielded very low strength across all the dosages.
- Mortar samples without spent wash exhibited 13.5 kN/m^2 flexural strength at 28 days, while those with 2% spent wash attained 15.3 kN/m^2. Strengths decreased with 4% and 6% spent wash compared to the 2% dosage. Spent wash stored for 6 months produced very low strength across all the dosages. The addition of fibers contributed to the strength improvement.

To conclude, while spent wash can influence the mortar strength, its impact is multifaceted and relies on various factors. Appropriate dosage, compatibility, and consideration of environmental conditions are crucial for leveraging its potential benefits while minimizing adverse effects on the mortar strength. Conducting experimental tests under specific project conditions would provide more precise insights into the impact of molasses on the mortar strength.

References

1. Z. Alwan, P. Jones, P. Holgate, Strategic sustainable development in the UK construction industry, through the framework for strategic sustainable development, using building information modelling. J. Clean. Prod. **140**, 349–358 (2017)
2. W. Ahmad, A. Ahmad, K.A. Ostrowski, F. Aslam, P. Joyklad, A scientometric review of waste material utilization in concrete for sustainable construction. Case Stud. Constr. Mater. **15**, e00683 (2021)
3. S. Kang, Y. Zhao, W. Wang, T. Zhang, T. Chen, H. Yi, F. Rao, S. Song, Removal of methylene blue from water with montmorillonitenanosheets/chitosan hydrogels as adsorbent. Appl. Surf. Sci. **448**, 203–211 (2018)
4. W. Wang, Y. Zhao, H. Bai, T. Zhang, V. Ibarra-Galvan, S. Song, Methylene blue removal from water using the hydrogel beads of poly (vinyl alcohol)-sodium alginate-chitosan-montmorillonite. Carbohydr. Polym. **198**, 518–528 (2018)
5. S. Mohana, B.K. Acharya, D. Madamwar, Distillery spent wash: treatment technologies and potential applications. J. Hazard. Mater. **163**, 12–25 (2009)
6. Y. Satyawali, M. Balakrishnan, Wastewater treatment in molasses-based alcohol distilleries for COD and color removal: a review. J. Environ. Manag. **86**, 481–497 (2008)
7. R. Chandra, R.N. Bharagava, V. Rai, Melanoidins as major colourant in sugarcane molasses-based distillery effluent and its degradation. Bioresour. Technol. **99**, 4648–4660 (2008)
8. C.V. Naik, A.S. Parlikar, A.B. Revade, Uses of spent wash to improve soil properties for road construction. Int. J. Civ. Eng. Technol. **9**(9), 1201 (2018)
9. P. Sharma, H. Joshi, Utilization of electrocoagulation treated spent wash sludge in making building blocks. Int. J. Environ. Sci. Technol. **13**, 349 (2015). https://doi.org/10.1007/s13762-015-0845-7
10. G.K. Arunvivek, R. Saravana Kumar, M. Senthil Kumar, M. Logesh Kumar, Experimental study on plasticizing effects of distillery spent wash in concrete. Int. J. Civ. Struct. Environ. Infrastruct. Eng. Res. Dev. **3**(5), 145–152 (2013). ISSN(P):2249-6866

11. M.U. Hassan, M. Aamer, M.U. Chattha, T. Haiying, I. Khan, Sugarcane distillery spent wash (DSW) as a bio-nutrient supplement: a win-win option for sustainable crop production. Agronomy **11**(1), 183 (2021). https://doi.org/10.3390/agronomy11010183
12. Y.S. Jagtap, S.B. Kankal, Investigation of optimum parameters for pollutant removal from distillery spent wash. Int. J. Eng. Res. Technol. **3**(7), 1566–1570 (2014). ISSN: 2278-0181
13. K. Kishore, Mix design for concrete roads as per IRC-15-2002. J. Build. Eng. (2010)
14. A.R. Djamaluddin, M.A. Caronge, M.W. Tjaronge, R. Irmawaty, Evaluation of sustainable concrete paving blocks incorporating processed waste tea ash. Case Stud. Constr. Mater. **12**, e00325 (2019). https://doi.org/10.1016/j.cscm.2019.e00325
15. S.K. Sindhu, A. Sharma, S. Ikram, Analysis and recommendation of agriculture use of distillery spent wash in Rampur district, India. J. Chem. **4**(3), 390–396 (2007). ISSN:0973-4945
16. R. Siddique, Utilization of industrial by-products in concrete. Procedia Eng. **95**, 335–347 (2014)
17. M.R. Bhagat, P.D. Biramane, C.V. Naik, Experimental investigation on effect of spent wash usage in mixing water on properties of concrete. Int. J. Res. Appl. Sci. Eng. Technol. **11**(6), 423 (2023)
18. IS: 269–2015 Ordinary Portland cement Specifications
19. ASTM C348–08. Standard test method for flexural strength of hydraulic-cement mortars (2008)
20. IS: 4031 (Part 6) 1988. Methods of physical tests for hydraulic cement, Part 6 determination of compressive strength of hydraulic cement other than masonry cement (1988)
21. ASTM C31/C31M 09. Standard practice for making and curing concrete test specimens in the field (2009)
22. ASTM C496/C496M-11. Standard test method for splitting tensile strength of cylindrical concrete specimens (2011)
23. IS 516:1999. Method of tests for strength of concrete (1999)

Sustainable Development with Earth Concrete

Mojtaba Kohandelnia, Masoud Hosseinpoor, and Ammar Yahia

1 Introduction

Cement, as the second most globally used material after water, is responsible for 8% of worldwide CO_2 emissions [1]. The main environmental drawback of cement and concrete production lies on their high energy demand, which leads to global warming and substantial consumption of natural resources [2]. Therefore, there is an urgent need for novel sustainable construction materials to reduce their negative environmental impact for different applications. Supplementary cementitious materials (SCMs) have been proposed as alternatives for this purpose to mitigate CO_2 emission [3]. To conserve natural resources while accommodating the growing demand for construction, earthen materials have demonstrated a great eco-friendly potential given their abundant local availability [4]. Traditional earth constructions have been historically used for different applications depending on the required mechanical performances [2].

Earth constructions include wattle and daub, cob, masonry adobe bricks, and rammed earth (RE), with reduced environmental impacts and embodied energy, good thermal performance, and diverse finished textures with/without stabilization [5–7]. These construction materials diminished over the last century due to the rapid emergence of new materials including cement-based mortars and concrete [8]. Earthen materials help minimize the consumption of new manufactured resources and utilization of recycled and locally sourced materials. Stabilization through cement and lime will enhance the mechanical performance and durability of earthen materials. The energy in construction and compaction is minimal compared to the high energy required for cement production in cement-stabilized rammed earth

M. Kohandelnia (✉) · M. Hosseinpoor · A. Yahia
Department of Civil and Building Engineering, Université de Sherbrooke, Sherbrooke, QC, Canada
e-mail: Mojtaba.Kohandelnia@usherbrooke.ca

A. Bahrami (ed.), *Advances in Sustainable Concrete for Construction*,
https://doi.org/10.1007/978-3-031-85052-3_9

(CSRE) [9]. Therefore, minimizing stabilizer content is extremely important to ensure that the final product is sustainable. Regardless of the offered advantages of earthen materials, their construction process is usually labor-intensive, time-consuming, and vulnerable to water, which can compromise their structural integrity.

Self-consolidating earth concrete (SCEC) offers a promising solution to address the aforementioned deficiencies with no compaction requirement [10, 11]. Clay and silt act as binder in earth composition, as well as sand and gravel as aggregates [10]. With an addition of a stabilizer, a blended binder system should be considered in SCEC systems [12]. Due to the variety of available soils, there can be diverse shapes, sizes, mineralogical properties, and clay fractions with very fine particles and high specific surface area (SSA), which can result in inconsistent performance [4, 13]. The presence of clay particles can impede the dispersion and flowability of SCEC systems since they result in higher yield stress and plastic viscosity values, owing to increased interparticle friction [14–17]. Achieving the required early-age compressive strength necessary for demolding remains an additional challenge. Selecting locally available soils based on their specific characteristics not only contributes to the design process but also helps predict the final performance of SCEC.

This chapter discusses the advantages and disadvantages of earth constructions, with their progress in being incorporated into concrete formulations. Given the benefits offered by earthen materials, this chapter investigates the advancement of designing SCEC and the existing challenges in its formulation.

2 Earth Construction

In recent times, there has been a noticeable surge in the interest surrounding the utilization of enhanced earth construction methods, driven by the desire to leverage their positive environmental attributes, energy conservation benefits, and thermal efficiency [18, 19]. A summary of the advantages and disadvantages of earthen materials is presented in Fig. 1. Moreover, it highlights the potential challenges to improve the workability of the earth-based concrete to achieve SCEC.

Soil composition varies in particle size, ranging from clay, silt, sand, to gravel, with dimensions that can be as small as sub-micrometers or several millimeters, depending on the classification scheme [4]. Strength primarily originates from factors such as particle interlocking, type and quantity of clay as binder, and pore water suction. Gerard et al. [20] revealed that water suction functions as an internal confining stress, which depends on relative humidity (RH) and voids. The balance between pore air pressure (u_a) and pore water pressure (u_w) can result in an additional stress acting as apparent cohesion ($s = u_a—u_w$) [21]. Cohesion increases with water suction due to the development of additional capillary cohesion [22]. The compaction process and drying cycle also contribute to strength development. Material drying enhances strength as it becomes unsaturated, with soil particles enveloped by air and water [23]. By cement stabilization, the main source of strength development derives from the hydration process, forming a solid matrix that binds

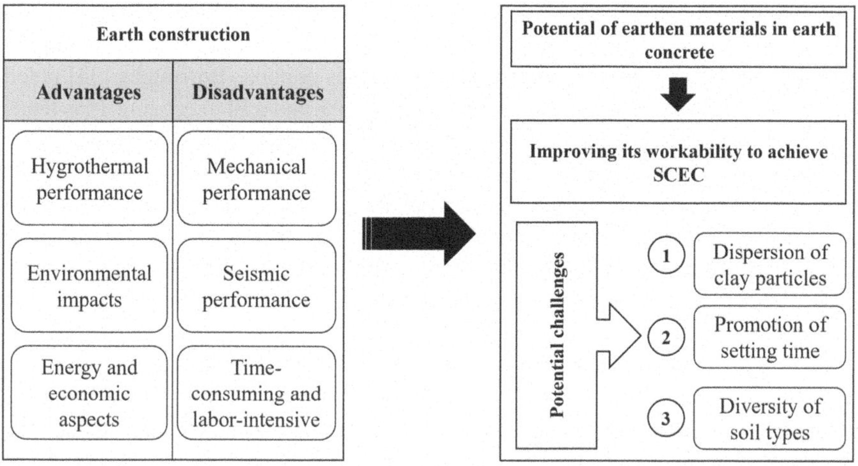

Fig. 1 Advantages and disadvantages of earth constructions and potential challenges to achieve SCEC

the soil particles [24, 25]. However, in lime-stabilized earth, the formation of cementing agents, primarily silica hydrate products, is responsible for imparting strength [26].

A diverse range of soils is suitable for earth construction. In general, a blend of sand and gravel, along with silt and clay as binding agents, is typically required [13]. However, dynamic compaction process can adapt larger particles into smaller ones, allowing a broader spectrum of soils to be suitable [27]. Bahar et al. [6] found that a high proportion of larger particles is suitable for RE. In certain instances, recycled materials may also be useful in earth construction [28]. It is crucial to emphasize proper particle packing to ensure increased density and strength. In the case of CSRE, the recommended contents of cement and finer clay particles (less than 2 μm) should fall within the ranges of 4% to 12% and 5% to 20% by mass of the total solid particles, respectively [13]. Perera and Jayasinghe [29] suggested that the mass fraction of fine constituents, including clay and silt, should remain below 30% to achieve greater strength. They also reported a sharp decline in strength when fine content surpasses 40% of the total solids. Concerning the composition of the earthen material, it is essential to control shrinkage. Notably, sand particles have been found effective in reducing shrinkage-induced deformations [6]. Conversely, the clay content significantly impacts strength development and shrinkage. Bryan [30] recommended maintaining low clay content for improved compaction while limiting it to reduce shrinkage and ensure effective stabilization. The types of clay minerals present in the earth noticeably influence swelling potential, ranging from high potential in montmorillonite/bentonite to very low in kaolinite [31]. This potential in earth compositions plays a crucial role in governing shrinkage over time.

An excessive clay content can result in undesirably high soil plasticity. The suitability of soil for construction is often assessed using its Atterberg limits [32], encompassing the liquid limit (LL) and plasticity index (PI), which characterize soil plasticity. Recommendations typically require LL and PI values below 45% and in the range of 10% to 30%, respectively [13]. In this context, Burroughs [33] noted that a soil with a linear shrinkage (LS) exceeding 6% and a PI exceeding 15% tends to exhibit a limited propensity for gaining strength. Another essential parameter for comprehending soil behavior is clay activity (A_c), defined as the ratio of PI to the percentage of clay particles finer than 2 μm. The specific value of A_c depends on the type of clay minerals present, and various levels of clay activity are outlined in Table 1. It is crucial to note that high A_c values are not favorable to earth constructions [34]. This relates to higher plasticity and, consequently, reduced strength and increased shrinkage.

A primary challenge in earth construction lies in its susceptibility to water, resulting in increased plasticity owing to clay content, which is unfavorable for strength development. In the early stages, effective curing is crucial to minimize the shrinkage of cement-stabilized soil. Jayasinghe and Kamaladasa [27] observed that shrinkage of cement-stabilized earth is particularly pronounced during the initial 4 days but slows down at later ages. For cement-stabilized earth, 28 days of curing at an RH of 95% and room temperature typically yields satisfactory mature compressive strength values. On the other hand, in lime-stabilized earth construction, the curing period may need to be extended, ranging from 7 to 60 weeks, to achieve maturity [36, 37].

The predominant uses of earth-based materials are observed in constructing building envelopes, typically for one- or two-story structures. These environmentally friendly materials can also serve as structural components, provided that their strength is augmented through stabilization while preserving their minimal environmental footprint. The sides of the walls tend to dry more quickly, whereas it may take a very long time, in some cases, for the core of the walls to dry completely. The rate of this hygroscopic transfer is primarily dictated by permeability, and the drying of the material can alter the hygroscopic conditions. François et al. [38] noted that the plastic hinges formed in the wall vanish when the wall dries, causing the wall's behavior to become more elastic. Additionally, suction increases with time, leading to higher strength development. It should also be noted that the interior side of the wall is less susceptible to water sensitivity compared to its exposed surfaces.

The inclusion of reinforcement in unstabilized earth construction is not recommended due to a lack of a robust anchorage [39]. The service life of reinforced elements depends on the onset and progression of corrosion. Nevertheless, the high alkalinity of cement paste can shield steel reinforcing bars from corrosion [40].

Table 1 Classification of A_c values [35]

Clay type	A_c (PI/%clay)
Low activity	$A_c < 0.75$
Medium activity	$0.75 < A_c < 1.25$
Active	$1.25 < A_c < 2.0$
High activity	$A_c > 2.0$

Meek et al. [41] indicated that the pH level of CSRE provides an appropriate environment for passivating steel reinforcement. Similar to conventional concrete, the hydrated cement particles contribute to the alkaline nature of the matrix. However, any decline in alkalinity can lead to steel depassivation, often brought about by carbonation or chloride exposure. Therefore, the moisture content of earth constructions plays a pivotal role in determining the potential for corrosion.

3 Advantages

Incorporation of earthen materials can offer a promising alternative to conventional cement concrete leading to reduced negative environmental impacts of the construction industry. Hygrothermal performance, environmental impacts, and energy and economic aspects of earth constructions are explored in this section to assess their potential for further applications.

3.1 Hygrothermal Performance

Energy required for heating, ventilation, and air-conditioning is affected by the hygrothermal characteristics of the materials in building envelopes [42]. Energy loss can be minimized with improved hygrothermal properties of materials quantified by heat and mass transfer through the porous medium [43].

3.1.1 Hygric Performance

Energy efficiency by heat transfer through the building envelope can be affected by the moisture presence. Therefore, it is important to resist the building envelope from moisture to reduce thermal transmittance, which leads to reduced energy consumption. Moisture buffering value and water vapor permeability are crucial measurements to understand uptake/release capacity and moisture transfer in building materials, respectively [44]. Earth brick samples exhibit high permeability values, ranging from 1.45 E-10 to 1.80 E-10 $kg.s^{-1}.m^{-1}.Pa^{-1}$ [45], while similar high permeability values have been reported for RE [46] and light earthen materials [47, 48].

Sorption isotherm, as another hygric performance which is related to the water content with respect to RH values, demonstrates the hygroscopic properties of the materials. Hygroscopic materials can help stabilize indoor RH variations, leading to an improvement of indoor air quality for dwellers [49]. Clay particles in earth have shown great hygroscopic properties, from the highest isotherm values related to montmorillonite clays followed by illite and kaolinite [50]. This can explain the superior indoor air quality of buildings with earthen materials compared to other industrial materials containing chemicals like heavy metal. It should be mentioned that high RH levels result in condensation, material deterioration, moisture accumulation, and fungal growth [46].

3.1.2 Thermal Performance

Thermal performance is affected by temperature and humidity, as well as material's characteristics including chemical composition, porosity, and microstructure [51]. Earth-based materials are well known for their high thermal mass, resulting in less temperature fluctuations compared to other materials [52]. RE walls have indicated low thermal conductivity with high heat capacity, which are favorable to maintain thermal stability in the residence [53]. According to the literature, the thermal conductivity of earth constructions, including RE and unfired earth, falls within the range of 650 to 1100 $mW.m^{-1}.K^{-1}$ [46, 54–56]. Notably, a 60-cm-thick cob wall was reported to exceed the required thermal stability as per the UK residential standards [57]. Straw and earth block masonry constructions have also been recognized for their commendable thermal performance in the United Kingdom [58–60]. Soebarto [61] observed comparable indoor temperatures for insulated and uninsulated RE structures during the summer, but noted a 5 °C lower temperature in the latter during the winter. It is worth noting that thermal conductivity increases with density, and the lowest values can be attributed to light earthen materials, followed by adobe, cob, wattle and daub, RE, and compressed earth blocks [44].

Thermal mass and insulation increase with increased wall thickness, and materials with higher thermal mass values tend to exhibit a slower rate of heat transfer. Soudani et al. [62] reported that temperature buffering and time lag (the difference between the highest interior and exterior temperatures) increase with higher thermal mass values. To gauge the thermal mass of a construction material, thermal diffusivity and effusivity properties are primarily employed. These properties show the capacity to convey temperature changes and its ability to interact with its surroundings and absorb heat, respectively. An effective strategy to enhance the thermal performance is to seek a balance between low diffusivity and high effusivity. Earthen materials (dry, wet, or solid bricks) present a balanced low diffusivity and high effusivity compared to other wall systems, highlighting their substantial thermal mass [62]. Materials with high thermal mass can mitigate indoor temperature variations in both winter and summer, which can save energy costs [63–65]. For instance, it has been reported that RE systems demonstrated limited interior temperature variations under varying exterior conditions [62].

3.2 Environmental Impacts

Earthen materials emerge as a promising solution for circular construction. Cabeza et al. [66] revealed that timber and earthen materials display the best performance in terms of embodied energy and CO_2 emissions. Comparative assessments have illustrated that unstabilized rammed earth (URE) exhibits lower environmental impacts than stabilized RE, highlighting the considerable ecological advantages of earth-based materials concerning ecosystems, human health, and resource conservation [67]. Leveraging locally sourced earth and introducing a minimal amount of stabilizing agents, such as cement, can contribute to reduced values of embodied carbon and energy [68].

Sustainable construction necessitates the adoption of novel production concepts, aligned in a circular model encompassing production, utilization, and recycling. The primary objective is to reduce waste generation and fresh resource consumption [69–71]. Notably, earth-based and vernacular construction techniques have emerged as promising alternatives to current building technologies. These construction techniques have been used for centuries, primarily owing to their reliance on locally available materials, minimal transportation requirements from extraction to the construction site, low embodied energy, efficient manufacturing processes, and reduced negative environmental impacts [68, 72].

3.3 Energy and Economic Aspects

The energy consumption comprises both embodied and maintenance energies, which are dependent on the geography of the building. The consumption of locally sourced earthen materials can, therefore, serve as a significant factor in reducing energy consumption. Venkatarama Reddy and Prasanna Kumar [9] reported that compaction energy increases with higher clay content and dry density. Nonetheless, it is essential to emphasize that the energy consumption related to the compaction process is fairly negligible compared to the energy content of cement. Therefore, it is recommended to identify the parameters that influence strength rather than increasing the cement content, which is associated with a noticeably high embodied energy. Increased dry density (lower porosity) and curing periods, as well as incorporating materials such as fly ash and lime, can contribute to achieving higher compressive strength values with lower negative environmental impacts [73].

Employing local earth as a construction material has the potential to lower construction costs by using locally available materials and the local labor force [69]. Earth-based constructions can also play a critical role in energy conservation and cost reduction associated with air-conditioning systems, given their appropriate hygrothermal properties. The combination of earth construction and insulating materials can deliver efficient thermal performance, consequently reducing the expenses related to cooling and heating energy [42]. Therefore, it can be concluded that earth-based construction offers substantial cost and energy savings in pre- and post-construction phases.

4 Disadvantages

Besides the advantages offered by earth construction, their drawbacks should be also well identified. These include the relatively poor mechanical and seismic performances of earth constructions, as well as their time- and energy-consuming construction processes [74].

4.1 Mechanical Properties

Earthen materials show varying strength characteristics depending on their moisture content, with higher strength in dry conditions and reduced strength in saturated states. The cohesion of earth is compromised when it contains higher water content, particularly if it includes a significant proportion of sand and coarser-sized silt fractions [75], which can promote the risk of de-cohesion. On the other hand, the hygroscopic conditions compromise mechanical performance of the earth [76]. Many empirical studies have presented that partially drying earthen materials lead to a substantial increase in their strength and stiffness [25, 77]. However, all earth types are not inherently suitable for achieving satisfactory strength. The compressive strength of optimized URE typically falls within the range of 0.5 to 2.5 MPa [78, 79]. Consequently, stabilization with cement or lime is often employed to enhance strength and reduce the chance of swelling, shrinkage, and cracking [80, 81]. Cement can be an effective stabilizer when the soil contains a higher proportion of sand and gravel, while soils with a greater silt and clay content are better stabilized with hydrated and natural lime [80].

Dao et al. [82] observed that the introduction of cement into clayey soil for adobe production led to the formation of calcite, C–S–H, and ettringite, which enhance resistance to water erosion by increasing compressive strength. However, the addition of 2% and 4% of cement by mass resulted in a decrease in thermal conductivity, while higher cement content caused an increase in thermal conductivity. This highlights the remarkable influence of the microstructure when cement is added. Van Damme and Houben [4] reported that using cement to stabilize bricks, RE, and clay concrete appears to be inefficient. This is because it necessitates nine, seven, and three times more cement content compared to high-performance concrete to achieve a 1 MPa compressive strength, respectively. Consequently, cement stabilization may not be the most suitable approach for earthen materials, except in cases where high strength values are not a requirement. Achieving optimal strength should be prioritized through alternative methods, such as optimizing the packing of the granular skeleton [28].

4.2 Seismic Performance

The seismic performance of structural earth-based elements is markedly limited, as has been frequently observed during various earthquakes worldwide [83–85]. The seismic performance of these structural elements is primarily influenced by the mass of the wall supports, which is a result of the vertical dead and live loads. This can also lead to the generation of significantly high inertial stresses [86]. Morris and Walker [87] examined the post-seismic performance of RE walls wherein only minor cracks were seen. This reflection confirms that RE buildings can exceed predicted performance expectations under seismic conditions. Bui et al. [88] stated that two-story RE buildings demonstrate satisfactory in-plane earthquake behavior up to ground motions of 0.23 g, as one-story RE buildings can resist motions up to 0.32 g without incurring significant damage. It has been reported that RE walls display greater resistance to seismic loading compared to adobe masonry walls, with values up to 40% higher [75]. This should be due to the homogeneity of RE compared to adobe masonry walls.

The layered structure of RE impacts the crack mechanism, but mechanical properties of RE do not appear to indicate distinct anisotropy [89]. Silva et al. [90] highlighted that traditional single-story RE buildings in Portugal with thick walls illustrated favorable seismic performance. However, the seismic performance of RE buildings requires further in-depth research. RE structures may not be suitable for regions with high risk of earthquake, owing to their limited compressive strength, low tensile strength, and high inertial forces (attributed to their large thickness and mass). Nevertheless, several post-seismic investigations have revealed that RE walls can exhibit satisfactory behavior [78, 79, 87, 91].

4.3 Time-Consuming and Labor-Intensive

The process of layer-by-layer compaction in RE, or compaction and installation of bricks, is identified for its extensive time and labor requirements. While this approach can be cost-effective in regions with affordable labor, it poses challenges in developed countries where labor costs are high. In addition, this system may give rise to crack formation between successive layers. Taking these limitations into account, a more fluid vertical earth concrete element with satisfactory thermal and environmental benefits emerges as a promising alternative to outperform traditional RE construction techniques.

5 Earth Concrete

There is the necessity to explore novel construction methods that harness the benefits of earth-based materials while advancing their utilization. Such an innovative material can be known as SCEC, which represents a new-generation approach, combining knowledge from self-consolidating concrete (SCC) and traditional earthen materials. There is an increasing interest in studies focusing on this emerging technique [10, 11, 92–94]. SCEC introduces the possibility of directly pouring the material into the molds instead of relying on the labor-intensive layer-by-layer mechanical compaction method. This not only enhances the surface quality of the end product but also reduces construction costs and time, contributing to the competitive advantages of using earth-based materials.

SCEC requires further consideration when compared to conventional concrete. This relies on the blended binder powder composition of SCEC, which includes fine clay and silt particles inherent in the earth along with the stabilizers used [10, 12, 95]. Aïsson et al. [96] reported that the rheological properties, stability, and compressive strength of SCC are notably influenced by the packing density and the quantity of fine particles. Moreover, the plastic viscosity and yield stress of the mortar matrix, serving as the suspending phase, tend to rise with an increase in the fine content, consequently leading to increased interparticle friction. This can enhance stability, but reduce flowability [15, 16]. As indicated in Table 2, earthen materials deliver good hygrothermal performance with minimal environmental impact. Nevertheless, their integration into SCEC is a tricky subject due to the substantial influence of cement and chemical admixtures, along with the formation of new microstructures. The following sections aim to explore the discussion of the three primary challenges associated with achieving SCEC.

Kohandelnia et al. [98] found that the sorption isotherm of SCEC was similar to that of clay bricks. However, SCEC showed higher water vapor diffusion resistance factors, falling within the range of 30–90, when compared to clay bricks. Figure 2 depicts the thermal diffusivity and effusivity of SCEC mixtures [98] compared to conventional building materials [62]. The results revealed that the effusivity of SCEC aligns with that of other earthen materials, while its diffusivity is lower than that of brick, dry earth, and wet earth. Nevertheless, SCEC maintains a favorable balance between low diffusivity and high effusivity when compared to other materials.

Figure 3 presents the efficiency of water, clay, and superplasticizer on yield stress and plastic viscosity of SCEC. As anticipated, an increase in water content results in a reduction of both yield stress and plastic viscosity, whereas higher clay content and fineness contribute to elevated rheological parameters. The Atterberg limits can be utilized as effective indicators for assessing the contribution of clay content and fineness in soil composition [12]. The study revealed that higher Atterberg limits are associated with more substantial increases in yield stress and plastic viscosity values. Furthermore, certain superplasticizers, such as polycarboxylate ether (PCE), polynaphthalene sulfonate (PNS), and linear polycarboxylate

Table 2 Comparing performance of earthen materials with SCEC

Characteristics	Earthen materials (RE and adobe)	SCEC
Hygrothermal performance	Various studies confirm their satisfying performance [53, 57, 97]	Comparable to clay bricks with respect to their microstructure [98]
Environmental impacts	Using locally natural materials with low impacts and embodied energy [68, 72]	Requires more research (stabilizers and admixtures can considerably affect)
Mechanical performance	Dependent on packing and stabilization [4, 28]	Dependent on stabilizer and water content as well as clay type/content [12, 99]
Time-consuming and labor-intensive	Yes	No
Layer-by-layer deficiencies	Yes	No
Complex architectural limitations	Yes	No
Shrinkage control	Controlled mostly by clay fraction (due to their sensitivity to water) [30, 100]	Dependent on early-age compressive strength and binder constituents [99]
Rheology	Not critical	Dependent on water content and clay type/content [17]
Dispersion of clay particles	Not critical	Challenge I—nucleation of fine particles and physicochemical interaction between clay and chemical admixtures
Formwork removal	Drying with passing time [25, 77]	Challenge II—promotion of setting time
Mix proportioning	Particle size distribution (PSD) and plasticity [13, 33]	Challenge III—representative soil parameters

(LPC), demonstrated promising potential in decreasing both yield stress and plastic viscosity values in the presence of cement in the matrix [17].

5.1 Challenge I: Dispersion of Clay Particles

The presence of various clay types and quantities in earth-based materials presents a formidable challenge in the development of SCEC. The high SSA, water absorption, promotion of nucleation, and the tendency for fine particles to flocculate can impede the attainment of the desired flowability in SCC [101]. In fact, achieving the necessary deflocculation of clay particles to enhance flowability and realize SCEC is a complex task, especially due to the questionable compatibility of these particles with chemical admixtures. This compatibility hinges on colloidal interactions and adhesion, which are influenced by the mineralogy and clay content.

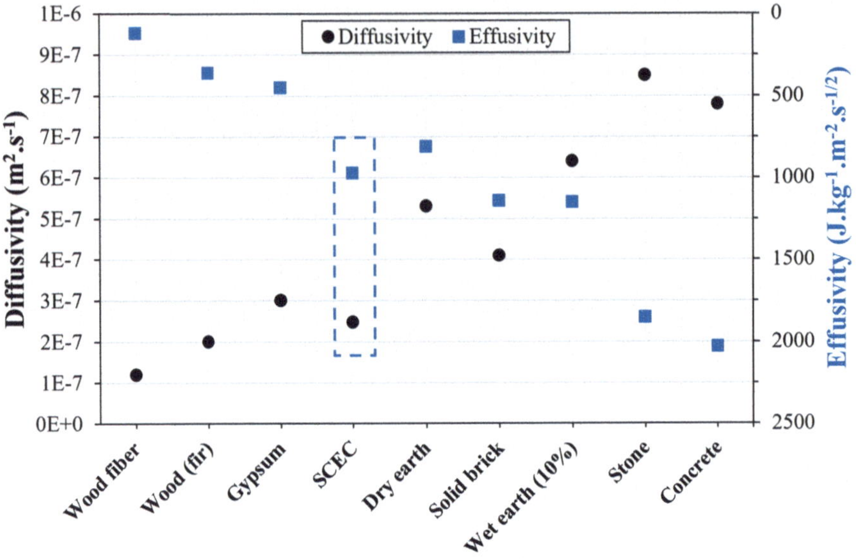

Fig. 2 Diffusivity and effusivity of SCEC compared to other conventional materials [62, 98]

Fig. 3 SCEC-rheology box: Changes in SCEC rheology by an increase in different constituents' contents

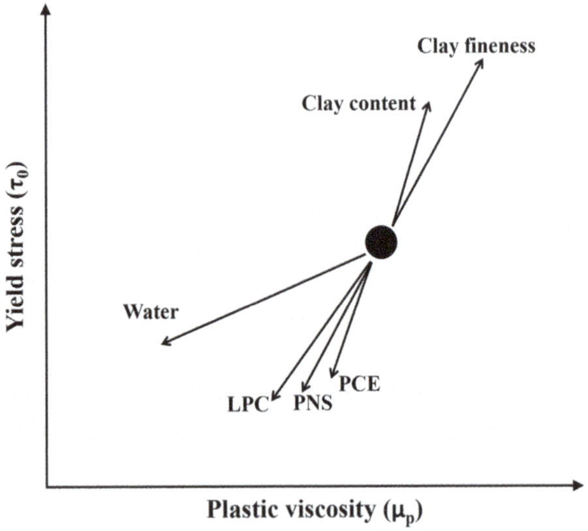

Figure 4 displays various clay minerals that serve as binding agents with notable swelling capabilities, which must be regulated to prevent potential failures. Among the common clay mineral types, kaolinite possesses a 1:1-layer structure, comprising aluminate and silicate groups connected via hydrogen bonds. These bonding forces result in non-swelling properties, which cause the dispersion and migration of kaolinite particles [102]. Meanwhile, 2:1-layer structure, such as

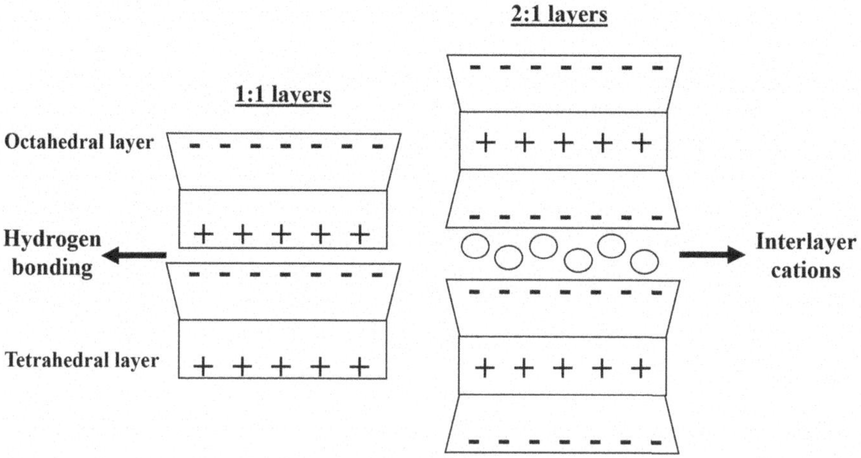

Fig. 4 Structures of 1:1 and 2:1 clay mineral

Table 3 CEC and SSA of clay minerals [31]

Mineral	Cation exchange capacity at pH 7 (milliequivalents per 100 g)	SSA (m^2/g)
Kaolinite	3–15	5–40
Halloysite (hydrated)	40–50	1100[a]
Illite	10–40	10–100
Chlorite	10–40	10–55
Vermiculite	100–150	760[a]
Smectite	80–120	40–800
Palygorskite–sepiolite	3–20	40–180
Allophane	30–135	2200
Imogolite	20–30	1540

[a]Depending on fraction of internal SSA

montmorillonite, illustrates a considerable degree of swelling. However, illite, also possessing a 2:1-layer structure, maintains a crystal structure stabilized by potassium bonding between its layers, leading to swelling capacities lower and higher than those of montmorillonite and kaolinite, respectively [103]. Cation exchange capacity (CEC), as a critical factor, is referred to the quantity of cations retained on particle surfaces, available for exchange with other cations at a given pH. Table 3 summarizes CEC and SSA of various clay minerals. It can be inferred that the clay content and type, as indicated by their CEC and SSA, have the potential to influence the performance of earth concrete through their interactions with dispersing agents.

Consequently, the compatibility between various chemical admixtures and the physical and chemical characteristics of clay within earthen materials remains uncertain in achieving SCEC. Ouellet-Plamondon et al. [10] introduced self-compacting clay-based concrete (SCCC) and observed that PCE has a notable impact on reducing yield stress of mixtures when a low cement content is present, thus enhancing their deformability. Furthermore, among commonly used deflocculants, sodium hexametaphosphate $(NaPO_3)_6$, referred to as NaHMP, demonstrated a significant dispersing effect on natural clay materials. These dispersing agents, often utilized in the cement and ceramic industries, can alter the surface interactions of clay particles. Their compositions lead to the adsorption of phosphate anions, consequently elevating the overall negative surface charge and alkalinity of the clay particles [104–107]. Sodium silicate (Na_2SiO_3), also represented as NaSil, has been widely employed to disperse raw clay and kaolin, exhibiting efficient deflocculant properties that increase the thickness of clay layers, accompanied by a pH rise of the solution [108–110].

Landrou et al. [93] evaluated the flowability of earthen materials without the inclusion of cement by utilizing NaHMP and NaSil as dispersants to create interparticle repulsive forces, thus deflocculating the clay particles. These dispersants showed a positive impact on enhancing the flowability of the earthen material. However, without cement, PCE did not contribute to reducing yield stress, possibly because PCE chains were adsorbed between the aluminosilicate layers, impeding their dispersibility [111, 112]. Zeta potential analysis confirmed the repulsive interaction between clay particles and dispersing agents. This affirmed that PCE made only minor changes due to its dispersing capability through steric hindrance, while the other clay dispersants induced dispersion through electrostatic repulsion [113]. Kohandelnia et al. [12] reported that high-range water reducer (HRWR) admixtures displayed a more positive impact on the dispersion of ternary binder system of SCEC systems (clay, silt, and cement) compared to clay dispersants. They found that LPC had the most noticeable effect in enhancing the flowability of SCEC mixtures, followed by PCE, PNS, NaHMP, and NaSil. Clay dispersants also led to higher viscosity values and a considerable increase in yield stress over time. This could be advantageous for 3D printing applications, as it improves shape stability, primarily owing to the presence of very fine clay particles [114]. 3D printing of earthen materials has gained increasing interest in recent years due to their customization and cost-efficiency [115–117].

5.2 *Challenge II: Promotion of Setting Time*

The early and late developments in compressive strength of SCEC play a pivotal role in determining the timing for formwork removal and ensuring long-term durability. Several studies explored the use of additional materials in conjunction with earth, such as coagulants, to expedite the demolding process. Landrou et al. [93] examined the effectiveness of calcium-based products in earthen materials and reported that different rates of Ca^{2+} release can lead to the precipitation of C–S–H phases in the presence of dispersing agents, exerting an anti-plasticizing effect. Additionally, they reported that the introduction of calcium products, especially

$Ca(OH)_2$, into the earth mixtures dispersed with NaHMP and NaSil, resulted in an increase in the storage modulus of the earthen suspensions over time. Scanning electron microscopy (SEM) images also revealed that the combination of NaSil and $Ca(OH)_2$ produced large calcium hydroxide crystal plates and typical C–S–H needles after 3 and 180 min, respectively. Another approach to expedite the demolding process involves the use of gelling agents, such as alginate [118, 119]. Chains of alginate, referred to alginic salts as seaweed biopolymers, create links by intercalating divalent cations [120], thereby enhancing the mechanical performance and resistance of the earthen material given their interaction with clay particles.

Ouellet-Plamondon et al. [10] formulated SCCC with a calcium sulfoaluminate cement (CSA) given its capacity to foster early-age strength. Achieving high early-age compressive strength is feasible through water reduction, owing to the use of compatible HRWR admixtures. The cement and water contents are the primary factors contributing to compressive strength development, followed by the interaction between chemical admixtures and binder constituents [12]. The introduction of SCMs, such as limestone filler and metakaolin, can also enhance early-age strength [121, 122].

5.3 Challenge III: Representative Soil Parameters Due to Diversity of Soil Types

Earth is a composite of particles characterized by diverse shapes, sizes, and mineralogical compositions [4]. The variation in mineral compositions across different extraction sites can lead to inconsistent performance of earthen materials. Therefore, predicting the behavior of a specific type of earth proves to be quite challenging due to the intricate nature of these materials and the numerous variables that must be taken into account [123]. Given the diversity of soils, it is fundamental to select soils with appropriate characteristics, depending on their clay content and type.

PSD, clay activity, and Atterberg limits serve as fundamental soil parameters essential for designing SCEC. Atterberg limits, which encompass the liquid and plasticity limits, offer insights into how soil behaves in the presence of water. PI is a fundamental soil characteristic, which can be correlated with other soil properties [124]. The extent of plasticity and cohesiveness can be inferred from PI. The plasticity of soil is significantly influenced by its clay type and content. Skempton [125] observed that PI of a soil increases with the clay fraction. Laskar and Pal [126] identified correlations between Atterberg limits and strength parameters. Voight [127] indicated the dependency of the residual shear strength of soil on its mineral composition. Atterberg limits can be employed to classify various soil types, and promising correlations have been established between workability of self-consolidating paste mixtures, Atterberg limits of the corresponding earth, and SSA of its ternary binder system [12].

Recommended considerations for the SCEC design are outlined in Table 4. The first critical factor is the type and content of clay, which can result in various soil characterization parameters for assessment. Ciancio et al. [128] proposed that a performance requirement for earth buildings, including RE, involves LL less than 45% and PI less than 30%. However, when LL and PI values exceed 40% and 25%, respectively, there may be a significant demand for admixtures to achieve SCEC [12]. The second parameter involves the type and content of cement used to expedite setting time and regulate durability. SCEC mixtures have achieved satisfactory strength with cement contents below 12% of the earth mass [12], and the utilization of other SCMs has the potential to reduce this further. The third parameter pertains to the type of admixture, which enables the dispersion of the ternary binder system in SCEC with different synergies. LPC, PCE, and PNS demonstrated better synergy with the ternary binder system of SCEC compared to NaHMP and NaSil [12]. Another important factor to consider is the paste volume, which affects the rheological behavior, workability, shrinkage, and mechanical performance of SCEC. Assuming the presence of a substantial fine content, including clay and silt, alongside cement and water in the paste results in a higher paste volume (exceeding 45%) compared to SCC [99]. The water-to-binder ratio (W/B) also impacts rheology by controlling the solid fraction, synergy of admixtures, and strength of SCEC. It can be concluded that using a compatible dispersing agent with a W/B ranging from 0.30 to 0.50 ensures the required workability of stable SCEC [10, 12, 93]. The packing of the granular skeleton should also be considered as a key factor influencing SCEC characteristics. Achieving an optimal packing density of aggregates can lead to a lower required paste volume and, consequently, a lower binder content to obtain the desired workability [129, 130]. Moreover, reduced

Table 4 Influencing parameters and recommended considerations to design SCEC

Influencing parameters	Contribution	Recommended considerations
1. Clay type/content	To evaluate different soil representative parameters	Desirability based on defining higher Atterberg limits and SSA (LL < 40% and PI < 25%)
2. Cement type/content	To promote setting time and controlling durability	Low enough to maintain low environmental impact (<12% of earth mass)
3. Chemical admixture type/dosage	To find optimum admixture type depending on powder constituents	PCE, PNS, and LPC showed better synergy compared to NaHMP and NaSil
4. Paste volume (V_P)	To assess the effect on rheology, workability, shrinkage, and mechanical performance	Higher than typical SCC due to the presence of ternary binder system (>45%)
5. Water-to-binder ratio (W/B)	To investigate the effect of water and solid fraction	Sufficiently low to attain greater strength and adequately high to enable dispersion (0.30–0.50)
6. Packing of granular skeleton	To minimize paste volume and binder content	In the range of available earth types (0.5–0.9)

interparticle voids in a well-packed aggregate skeleton decrease the water demand, thereby contributing to higher compressive strength [131, 132]. Based on the literature, the sand-to-total aggregate ratio between 0.5 and 0.9 lies in the range of aggregates present in earth [99]. Taking into account these influential parameters that impact workability, mechanical properties, and durability of SCEC, practical guidelines can be proposed for designing SCEC while considering soil characteristics.

6 Concluding Remarks and Perspectives

This chapter underscored the benefits of earth construction, which includes advantages in terms of thermal performance, energy efficiency, cost-effectiveness, and ecological considerations. While the numerous advantages of earth construction were highlighted, this study also discussed its drawbacks. Some of these limitations can be effectively mitigated through the incorporation of innovative construction techniques and the utilization of new value-added materials, such as SCEC. In light of this study, several conclusions and perspectives stand out:

- Exploring clay technology can enhance the compatibility of clay with admixtures, thereby improving construction outcomes. Selecting soil types based on their PSD and Atterberg limits is fundamental in achieving the desired workability and mechanical properties of SCEC.
- Coagulant agents, high early-age cementitious materials, and mix design optimization (including the packing and fine constituents) can result in improved mechanical and durability performance.
- The challenges of formulating SCEC were introduced, primarily for vertical elements, eliminating the need for compaction and offering versatile performance, making it well-suited for the prefabrication sector.
- To validate the concept, it is crucial to carry out comprehensive, large-scale comparative hygrothermal investigations in diverse climatic conditions. Furthermore, given the exceptional fire resistance of earthen materials, SCEC is expected to exhibit remarkable fire resistance, which should be further investigated.
- Due to the inclusion of various clay types and contents in SCEC mixtures, which have the ability to retain moisture and are responsive to temperature changes, it becomes essential to examine various curing methods through further research.

References

1. F. Winnefeld, A. Leemann, A. German, B. Lothenbach, CO2 storage in cement and concrete by mineral carbonation. Curr. Opin. Green Sustain. Chem. **38**, 100672 (2022). https://doi.org/10.1016/j.cogsc.2022.100672

2. F. Pacheco-Torgal, S. Jalali, Earth construction: Lessons from the past for future eco-efficient construction. Construct. Build Mater. **29**, 512–519 (2012). https://doi.org/10.1016/j.conbuildmat.2011.10.054
3. A. Coelho, Preliminary study for self-sufficiency of construction materials in a Portuguese region—Évora. J. Clean. Prod. **112**, 771–786 (2016). https://doi.org/10.1016/j.jclepro.2015.06.113
4. H. Van Damme, H. Houben, Earth concrete. Stabilization revisited. Cem. Concr. Res. **114**, 90–102 (2018). https://doi.org/10.1016/J.CEMCONRES.2017.02.035
5. H. Guillaud, H. Houben, *Earth Construction: A Comprehensive Guide* (Intermediate Technology Publications, London, 1994)
6. R. Bahar, M. Benazzoug, S. Kenai, Performance of compacted cement-stabilised soil. Cem. Concr. Compos. **26**, 811–820 (2004). https://doi.org/10.1016/j.cemconcomp.2004.01.003
7. M. Hall, Y. Djerbib, Rammed earth sample production: context, recommendations and consistency. Construct. Build Mater. **18**, 281–286 (2004). https://doi.org/10.1016/j.conbuildmat.2003.11.001
8. P.G. McHenry, *Adobe and Rammed Earth Buildings: Design and Construction* (University of Arizona Press, Tucson, AZ, 1989)
9. B.V. Venkatarama Reddy, P. Prasanna Kumar, Embodied energy in cement stabilised rammed earth walls. Energ. Buildings **42**, 380–385 (2010). https://doi.org/10.1016/j.enbuild.2009.10.005
10. C.M. Ouellet-Plamondon, G. Habert, Self-compacted clay based concrete (SCCC): proof-of-concept. J. Clean. Prod. **117**, 160–168 (2016). https://doi.org/10.1016/j.jclepro.2015.12.048
11. M. Kohandelnia, Development of Self-Consolidating Earth Concrete (SCEC) with Improved Multifunctional Performance for Green Construction, Université de Sherbrooke, 2022. http://hdl.handle.net/11143/20243
12. M. Kohandelnia, M. Hosseinpoor, A. Yahia, R. Belarbi, A new approach for proportioning self-consolidating earth paste (SCEP) using the Taguchi method. Construct. Build Mater. **347**, 128579 (2022). https://doi.org/10.1016/J.CONBUILDMAT.2022.128579
13. P. Walker, R. Keable, J. Martin, V. Maniatidis, *Rammed Earth: Design and Construction Guidelines* (IHS BRE, 2005)
14. M. Westerholm, B. Lagerbald, J. Silfwerbrand, E. Forssberg, Influence of fine aggregate characteristics on the rheological properties of mortars. Cem. Concr. Compos. **30**, 274–282 (2008)
15. S.K. Ling, A.K.H. Kwan, Adding ground sand to decrease paste volume, increase cohesiveness and improve passing ability of SCC. Construct. Build Mater. **84**, 46–53 (2015)
16. J.J. Chen, B.H. Li, P.L. Ng, A.K.H. Kwan, Adding granite polishing waste as sand replacement to improve packing density, rheology, strength and impermeability of mortar. Powder Technol. **364**, 404–415 (2020)
17. M. Kohandelnia, M. Hosseinpoor, A. Yahia, R. Belarbi, New insight on rheology of self-consolidating earth concrete (SCEC). Powder Technol. **424**, 118561 (2023). https://doi.org/10.1016/J.POWTEC.2023.118561
18. Scopus, Rammed earth (2023), https://www.scopus.com/term/analyzer.uri?sid=d663d876fa97225dc30b97b8706c15ef&origin=resultslist&src=s&s=TITLE-ABS-KEY%28rammed+earth%29&sort=plf-f&sdt=b&sot=b&sl=27&count=1294&analyzeResults=Analyze+results&txGid=3d2f51844a4157fd8990c2996effa619
19. Scopus, Earth construction (2022), https://www.scopus.com/term/analyzer.uri?sid=9f8fabd4c2c7b8cf8fe4311e0b0aaaca&origin=resultslist&src=s&s=TITLE-ABS-KEY%28Earth+Construction%29&sort=plf-f&sdt=b&sot=b&sl=33&count=21383&analyzeResults=Analyze+results&txGid=9a24ed872a46b718be86449c68237ecf
20. P. Gerard, M. Mahdad, A.R. Mccormack, B. François, A unified failure criterion for unstabilized rammed earth materials upon varying relative humidity conditions. Construct. Build Mater. **95**, 437–447 (2015). https://doi.org/10.1016/j.conbuildmat.2015.07.100
21. H. Nowamooz, C. Chazallon, Finite element modelling of a rammed earth wall. Construct. Build Mater. **25**, 2112–2121 (2011). https://doi.org/10.1016/j.conbuildmat.2010.11.021

22. P. Chauhan, A. El Hajjar, N. Prime, O. Plé, Unsaturated behavior of rammed earth: experimentation towards numerical modelling. Construct. Build Mater. **227**, 116646 (2019). https://doi.org/10.1016/j.conbuildmat.2019.08.027
23. P. Delage, Experimental unsaturated soil mechanics, in *Proceedings of the 3rd International Conference on Unsaturated Soils, Brazil*, (2002), pp. 973–996
24. C.T.S. Beckett, C.E. Augarde, *The Effect of Climate on the Unconfined Compressive Strength of Rammed Earth* (Elsevier, Amsterdam, 2011)
25. P.A. Jaquin, C. Augarde, D. Gallipoli, D.G. Toll, The strength of unstabilised rammed earth materials. Geotechnique **59**, 487–490 (2009)
26. M.R. Thompson, *Factors Influencing the Plasticity and Strength of Lime-Soil Mixtures, Tech. Rep. Bulletin 492* (Engineering Experiment Station, University of Illinios, Urbana, IL, 1967)
27. C. Jayasinghe, N. Kamaladasa, Compressive strength characteristics of cement stabilized rammed earth walls. Construct. Build Mater. **21**, 1971–1976 (2007). https://doi.org/10.1016/j.conbuildmat.2006.05.049
28. C. Jayasinghe, W.M.C.D.J. Fonseka, Y.M. Abeygunawardhene, Load bearing properties of composite masonry constructed with recycled building demolition waste and cement stabilized rammed earth. Construct. Build Mater. **102**, 471–477 (2016). https://doi.org/10.1016/j.conbuildmat.2015.10.136
29. A. Perera, C. Jayasinghe, Strength characteristics and structural design methods for compressed earth block walls. Mason. Int. **16**, 34–38 (2003)
30. A.J. Bryan, Criteria for the suitability of soil for cement stabilization. Build. Environ. **23**, 309–319 (1988)
31. N. Kumari, C. Mohan, Basics of clay minerals and their characteristic properties, in *Clay and Clay Minerals*, (IntechOpen, London, 2021)
32. ASTM D4318–17, Standard Test Methods for Liquid Limit, Plastic Limit, and Plasticity Index of Soils, ASTM Stand (2017)
33. S. Burroughs, Soil property criteria for rammed earth stabilization. J. Mater. Civ. Eng. **20**, 264–273 (2008). https://doi.org/10.1061/(ASCE)0899-1561(2008)20:3(264)
34. F. Pacheco-Torgal, S. Jalali, *Eco-Efficient Construction and Building Materials* (Springer Science & Business Media, Berlin, 2011)
35. P. Doat, A. Hays, H. Houben, S. Matuk, F. Vitoux, *Construire en terre* (Editions Alternative et Parallèles, Paris, 1979)
36. N. Consoli, P. Prietto, J. Carraro, K. Heineck, Behavior of compacted soil-Fly ash-carbide lime mixtures. J. Geotech. Geoenviron. Eng. **127**, 774–782 (2001)
37. J. Croft, The problem in predicting the suitability of soils for cementitious stabilization. Eng. Geol. **2**, 397–424 (1968)
38. B. François, L. Palazon, P. Gerard, Structural behaviour of unstabilized rammed earth constructions submitted to hygroscopic conditions. Construct. Build Mater. **155**, 164–175 (2017). https://doi.org/10.1016/j.conbuildmat.2017.08.012
39. P.J. Walker, S. Dobson, Pullout tests on deformed and plain Rebars in cement-stabilized rammed earth. J. Mater. Civ. Eng. **13**, 291–297 (2001)
40. K. Tuutti, Corrosion of Steel in Concrete, 1982
41. A.H. Meek, C.T.S. Beckett, M. Carsana, D. Ciancio, Corrosion protection of steel embedded in cement-stabilised rammed earth. Construct. Build Mater. **187**, 942–953 (2018). https://doi.org/10.1016/j.conbuildmat.2018.07.210
42. R.Y.M. Allam, *The Heat and Moisture Effect on the Structures* (La Rochelle University, La Rochelle, 2018)
43. O. Kaynakli, A review of the economical and optimum thermal insulation thickness for building applications. Renew. Sustain. Energy Rev. **16**, 415–425 (2012)
44. G. Giada, R. Caponetto, F. Nocera, Hygrothermal properties of raw earth materials: a literature review. Sustain. For. **11**, 5342 (2019). https://doi.org/10.3390/su11195342

45. R. Allam, N. Issaadi, R. Belarbi, M. El-Meligy, A. Altahrany, Hygrothermal behavior for a clay brick wall. Heat Mass Transf. **54**, 1579–1591 (2018). https://doi.org/10.1007/s00231-017-2271-5
46. M. Hall, D. Allinson, Analysis of the hygrothermal functional properties of stabilised rammed earth materials. Build. Environ. **44**, 1935–1942 (2009). https://doi.org/10.1016/J.BUILDENV.2009.01.007
47. F. Volhard, *Light Earth Building: A Handbook for Building with Wood and Earth* (Birkhauser, Basel, 2016)
48. M. Labat, C. Magniont, N. Oudhof, J.E. Aubert, From the experimental characterization of the hygrothermal properties of straw-clay mixtures to the numerical assessment of their buffering potential. Build. Environ. **97**, 69–81 (2016). https://doi.org/10.1016/J.BUILDENV.2015.12.004
49. M. Salonvaara, T. Ojanen, A. Holm, H.M. Künzel, A.N. Karagiozis, Moisture buffering effects on indoor air quality-experimental and simulation results, in *Proceedings of the Building IX, Clearwater, FL*, (2004)
50. M.N. Wuddivira, D.A. Robinson, I. Lebron, L. Bréchet, M. Atwell, S. De Caires, M. Oatham, S.B. Jones, H. Abdu, A.K. Verma, M. Tuller, Estimation of soil clay content from hygroscopic water content measurements. Soil Sci. Soc. Am. J. **76**, 1529–1535 (2012)
51. H. Zhang, Heat-insulating materials and sound-absorbing materials. Build. Mater. Civ. Eng., 304–423 (2011). https://doi.org/10.1533/9781845699567.304
52. A. Tavil, Thermal behavior of masonry walls in Istanbul. Construct. Build Mater. **18**, 111–118 (2004). https://doi.org/10.1016/J.CONBUILDMAT.2003.08.014
53. V.B. Lovec, M.Đ. Jovanovic–Popovic, B.D. Zivkovic, The thermal behavior of rammed earth wall in traditional house in Vojvodina: Thermal mass as a key element for thermal comfort. Therm. Sci. **2018**, 230 (2018)
54. M.L. Indekeu, M. Woloszyn, A.-C. Grillet, L. Soudani, A. Fabbri, Towards hygrothermal characterization of rammed earth with small-scale dynamic methods. Energy Procedia **132**, 297–302 (2017)
55. D. Allinson, M. Hall, Hygrothermal analysis of a stabilised rammed earth test building in the UK. Energ. Buildings **42**, 845–852 (2010). https://doi.org/10.1016/J.ENBUILD.2009.12.005
56. D. Medjelekh, L. Ulmet, F. Dubois, Characterization of hygrothermal transfers in the unfired earth. Energy Procedia **139**, 487–492 (2017). https://doi.org/10.1016/J.EGYPRO.2017.11.242
57. S. Goodhew, R. Griffiths, Sustainable earth walls to meet the building regulations. Energ. Buildings **37**, 451–459 (2005). https://doi.org/10.1016/J.ENBUILD.2004.08.005
58. C. MacDougall, Natural building materials in mainstream construction: lessons from the UK. J. Green Build. **3**, 3–14 (2008)
59. N. Desborough, S. Samant, Is straw a viable building material for housing in the United Kingdom? Sustain. For. **2**, 368–374 (2009)
60. C. Williams, S. Goodhew, R. Griffiths, L. Watson, The feasibility of earth block masonry for building sustainable walling in the United Kingdom. J. Build. Apprais. **6**, 99–108 (2010)
61. V. Soebarto, Analysis of indoor performance of houses using rammed earth walls, in *Eleventh International IBPSA Conference, Glasgow, Scotland*, (2009), pp. 1530–1537
62. L. Soudani, M. Woloszyn, A. Fabbri, J. Morel, A. Grillet, Energy evaluation of rammed earth walls using long term in-situ measurements. Sol. Energ. **141**, 70–80 (2017). https://doi.org/10.1016/j.solener.2016.11.002
63. N. Aste, A. Angelotti, M. Buzzetti, The influence of the external walls thermal inertia on the energy performance of well insulated buildings. Energ. Buildings **41**, 1181–1187 (2009). https://doi.org/10.1016/j.enbuild.2009.06.005
64. J.L. Fernandez, M.A. Porta-Gandara, N. Chargoy, Rapid on-site evaluation of thermal comfort through heatcapacity in buildings. Energ. Buildings **37**, 1205–1211 (2005)
65. P. Taylor, M.B. Luther, Evaluating rammed earth walls: a case study. Sol. Energ. **76**, 79–84 (2004). https://doi.org/10.1016/j.solener.2003.08.026

66. L.F. Cabeza, C. Barreneche, L. Miró, J.M. Morera, E. Bartolí, A. . Inés Fernández, Low carbon and low embodied energy materials in buildings: a review. Renew. Sustain. Energy Rev. **23**, 536–542 (2013). https://doi.org/10.1016/j.rser.2013.03.017
67. S. Serrano, C. Barreneche, L. Rincón, D. Boer, L.F. Cabeza, Optimization of three new compositions of stabilized rammed earth incorporating PCM: thermal properties characterization and LCA. Construct. Build Mater. **47**, 872–878 (2013). https://doi.org/10.1016/j.conbuildmat.2013.05.018
68. A. Arrigoni, C. Beckett, D. Ciancio, G. Dotelli, Life cycle analysis of environmental impact vs. durability of stabilised rammed earth. Construct. Build Mater. **142**, 128–136 (2017). https://doi.org/10.1016/j.conbuildmat.2017.03.066
69. J.C. Morel, A. Mesbah, M. Oggero, P. Walker, Building houses with local materials: means to drastically reduce the environmental impact of construction. Build. Environ. **36**, 1119–1126 (2001)
70. P.S. Ramesh, Energy Procedia appraisal of vernacular building materials and alternative technoligies for roofing and terracing options of embodied energy in buildings. Energy Procedia **14**, 1843–1848 (2012). https://doi.org/10.1016/j.egypro.2011.12.1177
71. G. Wadel, J. Avellaneda, A. Cuchí, Sustainability in industrialised architecture: closing the materials cycle. Inf. La Construcción **62**, 37–51 (2010)
72. J. Fernandes, R. Mateus, L. Bragança, The potential of vernacular materials to the sustainable building design, in *Vernacular Heritage and Earthen Architecture: Contributions for Sustainable Development*, (CRC, Vila Nova da Cerveira, 2013), pp. 623–629. https://doi.org/10.1201/b15685
73. C.G. da Rocha, N.C. Consoli, A.D.R. Johann, Greening stabilized rammed earth: Devising more sustainable dosages based on strength controlling equations. J. Clean. Prod. **66**, 19–26 (2014). https://doi.org/10.1016/j.jclepro.2013.11.041
74. M. Zami, A. Lee, Stabilised or un-stabilised earth construction for contemporary urban housing? in *5th International Conference on Responsive Manufacturing—Green Manufacturing (ICRM 2010), China*, (2010), pp. 11–13
75. L. Miccoli, U. Müller, P. Fontana, Mechanical behaviour of earthen materials: a comparison between earth block masonry, rammed earth and cob. Construct. Build Mater. **61**, 327–339 (2014). https://doi.org/10.1016/j.conbuildmat.2014.03.009
76. D. Toll, A framework for unsaturated soil behavior. Géotechnique **40**, 31–44 (1990)
77. F. Champiré, A. Fabbri, J. Morel, H. Wong, F. Mcgregor, Impact of relative humidity on the mechanical behavior of compacted earth as a building material. Construct. Build Mater. **110**, 70–78 (2016). https://doi.org/10.1016/j.conbuildmat.2016.01.027
78. V. Maniatidis, P. Walker, Structural capacity of rammed earth in compression. J. Mater. Civ. Eng. **20**, 230–238 (2008)
79. Q.-B. Bui, J.-C. Morel, S. Hans, N. Meunier, Compression behaviour of non-industrial materials in civil engineering by three scale experiments: the case of rammed earth. Mater. Struct. **42**, 1101–1116 (2009). https://doi.org/10.1617/s11527-008-9446-y
80. G. Minke, *Building with Earth: Design and Technology of a Sustainable Architecture* (Birkhäuser, Basel, 2006)
81. M.I. Gomes, T.D. Gonçalves, P. Faria, Hydric behavior of earth materials and the effects of their stabilization with cement or lime: study on repair mortars for historical rammed earth structures. J. Mater. Civ. Eng. **28**, 04016041 (2016)
82. K. Dao, M. Ouedraogo, Y. Millogo, J.E. Aubert, M. Gomina, Thermal, hydric and mechanical behaviours of adobes stabilized with cement. Construct. Build Mater. **158**, 84–96 (2018). https://doi.org/10.1016/J.CONBUILDMAT.2017.10.001
83. D. D'Ayala, G. Benzoni, Historic and traditional structures during the 2010 Chile earthquake: observations, codes, and conservation strategies. Earthq. Spectra **28**, 425–451 (2012)
84. M. Idália, M. Lopes, J. De Brito, Seismic resistance of earth construction in Portugal. Eng. Struct. **33**, 932–941 (2011). https://doi.org/10.1016/j.engstruct.2010.12.014

85. K. Sharma, L. Deng, C.C. Noguez, Field investigation on the performance of building structures during the April 25, 2015, Gorkha earthquake in Nepal. Eng. Struct. **121**, 61–74 (2016). https://doi.org/10.1016/j.engstruct.2016.04.043
86. Q.B. Bui, T.-T. Bui, A. Limam, Assessing the seismic performance of rammed earth walls by using discrete elements. Cogent Eng. **3**, 1200835 (2016)
87. H. Morris, R. Walker, Observation of the performance of earth buildings following the February 2011 Christchurch Earthquake. Bull. N. Z. Soc. Earthq. Eng. **44**, 358–367 (2011)
88. Q. Bui, A. Limam, T. Bui, Dynamic discrete element modelling for seismic assessment of rammed earth walls. Eng. Struct. **175**, 690–699 (2018). https://doi.org/10.1016/j.engstruct.2018.08.084
89. P.A. Jaquin, Analysis of Historic Rammed Earth Construction, Durham theses, Durham University, 2008. http://etheses.dur.ac.uk/2169/
90. R.A. Silva, N. Mendes, D.V. Oliveira, A. Romanazzi, O. Domínguez-martínez, T. Miranda, Evaluating the seismic behaviour of rammed earth buildings from Portugal: from simple tools to advanced approaches. Eng. Struct. **157**, 144–156 (2018). https://doi.org/10.1016/j.engstruct.2017.12.021
91. D. Ciancio, C. Augarde, Capacity of unreinforced rammed earth walls subject to lateral wind force: elastic analysis versus ultimate strength analysis. Mater. Struct. **46**, 1569–1585 (2013). https://doi.org/10.1617/s11527-012-9998-8
92. C. Ma, L. Chen, B. Chen, Experimental study of effect of Fly ash on self-compacting rammed earth construction stabilized with cement-based composites. J. Mater. Civ. Eng. **28**, 04016022 (2016). https://doi.org/10.1061/(ASCE)MT.1943-5533.0001518
93. G. Landrou, C. Brumaud, F. Winnefeld, R.J. Flatt, G. Habert, Lime as an anti-plasticizer for self-compacting clay concrete. Materials **9**, 330 (2016). https://doi.org/10.3390/ma9050330
94. M. Kohandelnia, M. Hosseinpoor, A. Yahia, R. Belarbi, Introduction of self-consolidating earth concrete (SCEC) and investigation on the influencing parameters using Taguchi method, in *First International RILEM Conference on Earthen Construction—Fundamentals and Applications for a Circular Economy, Paris, France*, (2022)
95. D. Gélard, *Identification and Characterization of Internal Cohesion of Earth Material in Its Natural Condition (in French)* (INPG Grenoble, Saint-Martin-d'Hères, 2005)
96. B.M. Aïssoun, S.-D. Hwang, K.H. Khayat, Influence of aggregate characteristics on workability of superworkable concrete. Mater. Struct. **49**, 597–609 (2016)
97. J. Pinto, A. Paiva, H. Varum, A. Costa, D. Cruz, S. Pereira, L. Fernandes, P. Tavares, J. Agarwal, Corn's cob as a potential ecological thermal insulation material. Energ. Buildings **43**, 1985–1990 (2011). https://doi.org/10.1016/J.ENBUILD.2011.04.004
98. M. Kohandelnia, M. Hosseinpoor, A. Yahia, R. Belarbi, Hygrothermal and microstructural characterization of self-consolidating earth concrete (SCEC). J. Build. Eng. **69**, 106287 (2023). https://doi.org/10.1016/J.JOBE.2023.106287
99. M. Kohandelnia, M. Hosseinpoor, A. Yahia, R. Belarbi, Multiscale investigation of self-consolidating earthen materials using a novel concrete-equivalent mortar approach. Construct. Build Mater. **370**, 130700 (2023). https://doi.org/10.1016/J.CONBUILDMAT.2023.130700
100. R. Bahar, M. Benazzoung, S. Kenai, Performance of compacted cement stabilized soil. Cem. Concr. Compos. **26**, 811–820 (2004)
101. Y. Qian, G. De Schutter, Enhancing thixotropy of fresh cement pastes with nanoclay in presence of polycarboxylate ether superplasticizer (PCE). Cem. Concr. Res. **111**, 15–22 (2018). https://doi.org/10.1016/J.CEMCONRES.2018.06.013
102. K.C. Lombardi, Structural and morphological characterization of the PP-0559 kaolinite from the Brazilian Amazon region. J. Braz. Chem. Soc. **13**, 270 (2002)
103. N. Mungan, Discussion of an overview of formation damage. J. Petrol. Tech. **41**, 1224 (1989)
104. K.O. Adebowale, I.E. Unuabonah, B.I. Olu-Owolabi, Adsorption of some heavy metal ions on sulfate- and phosphate-modified kaolin. Appl. Clay Sci. **29**, 145–148 (2005). https://doi.org/10.1016/j.clay.2004.10.003

105. F. Andreola, E. Castellini, T. Manfredini, M. Romagnoli, The role of sodium hexametaphosphate in the dissolution process of kaolinite and kaolin. J. Eur. Ceram. Soc. **24**, 2113–2124 (2004). https://doi.org/10.1016/S0955-2219(03)00366-2
106. F. Andreola, E. Castellini, J.M.F. Ferreira, S. Olhero, M. Romagnoli, Effect of sodium hexametaphosphate and ageing on the rheological behaviour of kaolin dispersions. Appl. Clay Sci. **31**, 56–64 (2006). https://doi.org/10.1016/j.clay.2005.08.004
107. F. Andreola, E. Castellini, G. Lusvardi, L. Menabue, M. Romagnoli, Release of ions from kaolinite dispersed in deflocculant solutions. Appl. Clay Sci. **36**, 271–278 (2007). https://doi.org/10.1016/j.clay.2006.10.002
108. J.L. Amorós, V. Beltrán, V. Sanz, J.C. Jarque, Electrokinetic and rheological properties of highly concentrated kaolin dispersions: influence of particle volume fraction and dispersant concentration. Appl. Clay Sci. **49**, 33–43 (2010). https://doi.org/10.1016/j.clay.2010.03.020
109. H.M.M. Diz, B. Rand, The mechanism of deflocculation of kaolinite by polyanions. Br. Ceram. Trans. **89**, 77–82 (1990)
110. D. Penner, G. Lagaly, Influence of anions on the rheological properties of clay mineral dispersions D. Appl. Clay Sci. **19**, 131–142 (2001)
111. L. Lei, J. Plank, A study on the impact of different clay minerals on the dispersing force of conventional and modified vinyl ether based polycarboxylate superplasticizers. Cem. Concr. Res. **60**, 1–10 (2014)
112. S. Ng, J. Plank, Interaction mechanisms between Na montmorillonite clay and MPEG-based polycarboxylate superplasticizers. Cem. Concr. Res. **378**, 222–231 (2012)
113. D. Marchon, S. Mantellato, A.B. Eberhardt, R.J. Flatt, Adsorption of chemical admixtures, in *Science and Technology of Concrete Admixtures*, (2016), pp. 219–256
114. S. Kawashima, K. Wang, R.D. Ferron, J.H. Kim, N. Tregger, S. Shah, A review of the effect of nanoclays on the fresh and hardened properties of cement-based materials. Cem. Concr. Res. **147**, 106502 (2021). https://doi.org/10.1016/J.CEMCONRES.2021.106502
115. A. Perrot, D. Rangeard, E. Courteille, 3D printing of earth-based materials: processing aspects. Construct. Build Mater. **172**, 670–676 (2018). https://doi.org/10.1016/J.CONBUILDMAT.2018.04.017
116. H. Alhumayani, M. Gomaa, V. Soebarto, W. Jabi, Environmental assessment of large-scale 3D printing in construction: a comparative study between cob and concrete. J. Clean. Prod. **270**, 122463 (2020). https://doi.org/10.1016/J.JCLEPRO.2020.122463
117. M. Gomaa, W. Jabi, A. Veliz Reyes, V. Soebarto, 3D printing system for earth-based construction: case study of cob. Autom. Constr. **124**, 103577 (2021). https://doi.org/10.1016/J.AUTCON.2021.103577
118. C. Galán-Marín, C. Rivera-Gómez, J. Petric, Clay-based composite stabilized with natural polymer and fibre. Construct. Build Mater. **24**, 1462–1468 (2010). https://doi.org/10.1016/J.CONBUILDMAT.2010.01.008
119. C.A. Dove, F.F. Bradley, S.V. Patwardhan, Seaweed biopolymers as additives for unfired clay bricks. Mater. Struct. **49**, 4463–4482 (2016)
120. T. Funami, Y. Fang, S. Noda, S. Ishihara, M. Nakauma, K.I. Draget, K. Nishinari, G.O. Phillips, Rheological properties of sodium alginate in an aqueous system during gelation in relation to supermolecular structures and Ca2+ binding. Food Hydrocoll. **23**, 1746–1755 (2009). https://doi.org/10.1016/J.FOODHYD.2009.02.014
121. A.M. Matos, H. Varum, Self-compacting earth-based composites: mixture design and multi-performance characterisation. Buildings **12**, 612 (2022). https://doi.org/10.3390/buildings12050612
122. M. Kohandelnia, A. Yahia, Rheology of stabilized earth-based paste. ACI Symp. Publ. **362**, 108–126 (2024). https://doi.org/10.14359/51740878
123. S. Guiheneuf, *Formulation et renforts de blocs en matériau terre pour une utilisation structurelle* (INSA de Rennes, Rennes, 2020) https://tel.archives-ouvertes.fr/tel-03194559
124. P.P. Raj, *Soil Mechanics and Foundation Engineering* (Dorling Kindersley, New Delhi, 2012)

125. A.W. Skempton, The Colloidal activity of clays, in *Proceedings of the Third International Conference on Soil Mechanics and Foundation Engineering, London*, (1953), pp. 47–61
126. A. Laskar, S.K. Pal, Geotechnical characteristics of two different soils and their mixture and relationships between parameters. Electron. J. Geotech. Eng. **17**, 2821–2832 (2012)
127. B. Voight, Correlation between Atterberg plasticity limits and residual shear strength of natural soils. Géotechnique **23**, 265–267 (1973)
128. D. Ciancio, P. Jaquin, P. Walker, Advances on the assessment of soil suitability for rammed earth. Construct. Build Mater. **42**, 40–47 (2013). https://doi.org/10.1016/j.conbuildmat.2012.12.049
129. X. Wang, K. Wang, P. Taylor, G. Morcous, Assessing particle packing based self-consolidating concrete mix design method. Construct. Build Mater. **70**, 439–452 (2014)
130. W. Zuo, J. Liu, Q. Tian, W. Xu, W. She, P. Feng, C. Miao, Optimum design of low-binder self-compacting concrete based on particle packing theories. Construct. Build Mater. **163**, 938–948 (2018)
131. S.A.A.M. Fennis, J.C. Walraven, J.A. den Uijl, Defined-performance design of ecological concrete. Mater. Struct. **46**, 639–650 (2013)
132. L.G. Li, J.J. Chen, A.K.H. Kwan, Roles of packing density and water film thickness in strength and durability of limestone fines concrete. Mag. Concr. Res. **69**, 595–605 (2017)

Index

A. Bahrami (ed.), *Advances in Sustainable Concrete for Construction*,
https://doi.org/10.1007/978-3-031-85052-3

The manufacturer's authorised representative in the EU is Springer Nature Customer Service Centre GmbH, Europaplatz 3, 69115 Heidelberg, Germany. If you have any concerns regarding our products, please contact ProductSafety@springernature.com

Printed and bound by CPI Group (UK) Ltd, Croydon, CR0 4YY
01/10/2026
02225850-0004